The Science of Soccer

John Wesson

IoP

Institute of Physics Publishing
Bristol and Philadelphia

British Library Cataloguing-in-Publication Data

A catalogue record of this book is available from the British Library.

ISBN 0 7503 0813 3

Library of Congress Cataloging-in-Publication Data are available

Commissioning Editor: John Navas
Production Editor: Simon Laurenson
Production Control: Sarah Plenty
Cover Design: Frédérique Swist
Marketing: Nicola Newey and Verity Cooke

Published by Institute of Physics Publishing, wholly owned by The Institute of Physics, London

Institute of Physics, Dirac House, Temple Back, Bristol BS1 6BE, UK

US Office: Institute of Physics Publishing, The Public Ledger Building, Suite 1035, 150 South Independence Mall West, Philadelphia, PA 19106, USA

Typeset by Academic + Technical Typesetting, Bristol
Printed in the UK by MPG Books Ltd, Bodmin, Cornwall

For Olive
My favourite football fan

Contents

Preface

Football is by far the world's most popular game. Millions play the game and hundreds of millions are entertained by it, either at football grounds or through television. Despite this the scientific aspects of the game have hardly been recognised, let alone discussed and analysed. This is in contrast to some other games which have received much more attention, particularly so in the case of golf.

What is meant by 'science' in the context of football? This book deals basically with two types of subject. The first is the 'hard science', which mainly involves using physics to uncover basic facts about the game. This ranges from understanding the comparatively simple mechanics of the kick to the remarkably complex fluid dynamics associated with the flight of the ball. The second group of subjects is diverse. There is the role of chance in deciding results and, more significantly, in influencing which team wins the Championship or the Cup. Is the winning team the best team? We look at the players and ask how their success varies with age. We also ask, what is the best height for footballers and, with almost incredible results, what is the best time of year for them to be born? Further subjects include analysis of the laws, various theoretical aspects of the play, and the economics of the professional game.

In the first nine chapters of the book these subjects are described without the use of mathematics. The mathematical analysis which underlies this description is saved for the tenth and final chapter. Most of the material in the book is original and in many areas the author has made progress only with the assistance of others. I must thank David Goodall for the help he gave in experiments on the bounce and flight of the ball, and both him and Chris Lowry for the experiments which produced the drag curve for a football. The on-field experiments were carried out with the help of Mickey Lewis and the Oxford United Youth team. My understanding of the development of the ball was much improved in discussions with Duncan Anderson of Mitre, and I have taken the information on club finances from the *Annual Review of Football Finance* produced by Deloitte and Touche.

I am grateful to John Navas, the Commissioning Editor at Institute of Physics Publishing. Without his interest and encouragement this book would not have seen the light of day. Thanks are also due to Jack Connor and John Hardwick who read the manuscript and made many helpful suggestions. The book uses, and depends upon, a large number of figures. These were all produced by Stuart Morris. I am very grateful to him for his skill and unfailing helpfulness. Finally, I must thank Lynda Lee for her care and dedication in typing the manuscript and dealing with the many corrections and re-writes this involved

John Wesson
January 2002

Chapter 1

1

The ball and the bounce

The ball

Ball-like objects must have been kicked competitively for thousands of years. It doesn't require much imagination to picture a boy kicking a stone and being challenged for possession by his friends. However the success of 'soccer' was dependent on the introduction of the modern ball with its well-chosen size, weight and bounce characteristics.

When soccer was invented in the nineteenth century the ball consisted of an ox or pig bladder encased in leather. The bladder was pumped through a gap in the leather casing, and when the ball was fully pumped this gap was closed with lacing. While this structure was a great advance, a good shape was dependent on careful manufacture and was often lost with use. The animal bladder was soon replaced by a rubber 'bladder' but the use of leather persisted until the 1960s.

The principal deficiency of leather as a casing material was that it absorbed water. When this was combined with its tendency to collect mud the weight of the ball could be doubled. Many of us can recollect the sight of such a ball with its exposed lacing hurtling toward us and expecting to be headed.

The period up to the late 1980s saw the introduction of multi-layer casing and the development of a totally synthetic ball. Synthetic fibre layers are covered with a smooth polymer surface material and the ball is inflated with a latex bladder. This ball resists the retention of water and reliably maintains its shape.

The casing of high quality balls is made up of panels. These panels, which can have a variety of shapes, are stitched together through pre-punched stitch holes using threads which are waxed for improved water resistance. This can require up to 2000 stitches. The lacing is long gone, the ball now being pumped through a tiny hole in the casing. Such balls are close to ideal.

The general requirements for the ball are fairly obvious. The ball mustn't be too heavy to kick, or so light that it is blown about, or will not carry. It shouldn't be too large to manoeuvre or too small to control, and the best diameter, fixed in 1872, turned out to be about the size of the foot. The optimisation took place by trial and error and the present ball is defined quite closely by the laws of the game.

The laws state that 'The circumference shall not be more than 28 inches and not less than 27 inches. The weight of the ball shall be not more than 16 ounces and not less than 14 ounces. The pressure shall be equal to 0.6 to 1.1 atmosphere.' Since 1 atmosphere is 14.7 pounds per square inch this pressure range corresponds to 8.8 to 16.2 pounds per square inch. (The usually quoted 8.5 to 15.6 pounds per square inch results from the use of an inaccurate conversion factor.)

From a scientific point of view the requirement that the pressure should be so low is amusing. Any attempt to reduce the pressure in the ball below one atmosphere would make it collapse. Even at a pressure of 1.1 atmosphere the ball would be a rather floppy object. What the rule really calls for, of course, is a pressure *difference* between the inside and the outside of the ball, the pressure inside being equal to 1.6 to 2.1 atmosphere.

Calculation of the ball's behaviour involves the mass of the ball. For our purposes mass is simply related to weight. The weight of an object of given mass is just the force exerted on that mass by gravity. The names used for the two quantities are rather confusing, a mass of one pound being said to have a weight of one pound. However, this need not trouble us; suffice it to say that the football has a mass of between 0.875 and 1.0 pound or 0.40 and 0.45 kilogram.

Although it will not enter our analysis of the behaviour of the ball, it is of interest to know how the pressure operates. The air in the atmosphere consists of very small particles called molecules. A hundred thousand air molecules placed sided by side would measure the same as the diameter of a human hair. In reality the molecules are randomly distributed in space. The number of molecules is enormous, there being 400 million million million (4×10^{20}) molecules in each inch cube. Nevertheless most of the space is empty, the molecules occupying about a thousandth of the volume.

The molecules are not stationary. They move with a speed greater than that of a jumbo jet. The individual molecules move in random directions with speeds around a thousand miles per hour. As a result of this motion the molecules are continually colliding with each other. The molecules which are adjacent to the casing of the ball also collide with the casing and it is this bombardment of the casing which provides the pressure on its surface and gives the ball its stiffness.

The air molecules inside the ball have the same speed as those outside, and the extra pressure inside the ball arises because there are more molecules in a given volume. This was the purpose of pumping the ball – to introduce the extra molecules. Thus the outward pressure on the casing of the ball comes from the larger number of molecules impinging on the inner surface as compared with the number on the outer surface.

The bounce

The bounce seems so natural that the need for an explanation might not be apparent. When solid balls bounce it is the

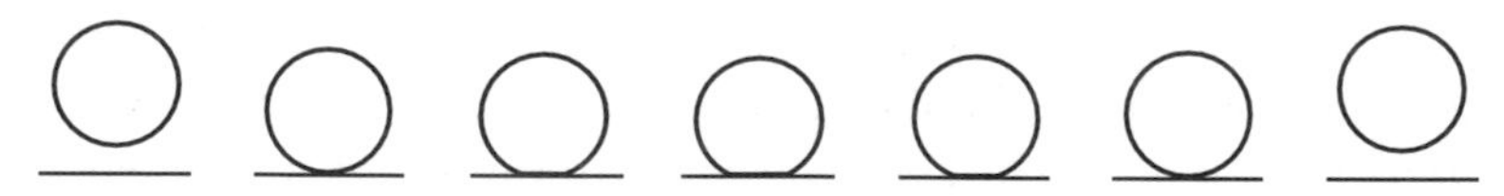

Figure 1.1. Sequence of states of the ball during the bounce.

elasticity of the material of the ball which allows the bounce. This applies for example to golf and squash balls. But the casing of a football provides practically no elasticity. If an unpumped ball is dropped it stays dead on the ground.

It is the higher pressure air in the ball which gives it its elasticity and produces the bounce. It also makes the ball responsive to the kick. The ball actually bounces from the foot, and this allows a well-struck ball to travel at a speed of over 80 miles per hour. Furthermore, a headed ball obviously depends upon a bounce from the forehead. We shall examine these subjects later, but first let us look at a simpler matter, the bounce itself.

We shall analyse the mechanics of the bounce to see what forces are involved and will find that the duration of the bounce is determined simply by the three rules specifying the size, weight and pressure. The basic geometry of the bounce is illustrated in figure 1.1. The individual drawings show the state of the ball during a vertical bounce. After the ball makes contact with the ground an increasing area of the casing is flattened against the ground until the ball is brought to rest. The velocity of the ball is then reversed. As the ball rises the contact area reduces and finally the ball leaves the ground.

It might be expected that the pressure changes arising from the deformation of the ball are important for the bounce but this is not so. To clarify this we will first examine the pressure changes which do occur.

Pressure changes

It is obvious that before contact with the ground the air pressure is uniform throughout the ball. When contact occurs and the bottom of the ball is flattened, the deformation increases the pressure around the flattened region. However, this pressure increase is rapidly redistributed over the whole of the ball. The speed with which this redistribution occurs is the speed of sound, around 770 miles per hour. This means that sound travels across the ball in about a thousandth of a second and this is fast enough to maintain an almost equal pressure throughout the ball during the bounce.

Although the pressure remains essentially uniform inside the ball the pressure itself will actually increase. This is because the flattening at the bottom of the ball reduces the volume occupied by the air, in other words the air is compressed. The resulting pressure increase depends on the speed of the ball before the bounce. A ball reaching the ground at 20 miles per hour is deformed by about an inch and this gives a pressure increase of only 5%. Such small pressure changes inside the ball can be neglected in understanding the mechanism of the bounce. So what does cause the bounce and what is the timescale?

Mechanism of the bounce

While the ball is undeformed the pressure on any part of the inner surface is balanced by an equal pressure on the opposite facing part of the surface as illustrated in figure 1.2. Consequently, as expected, there is no resultant force on the ball. However, when the ball is in contact with the ground additional forces comes into play. The casing exerts a pressure

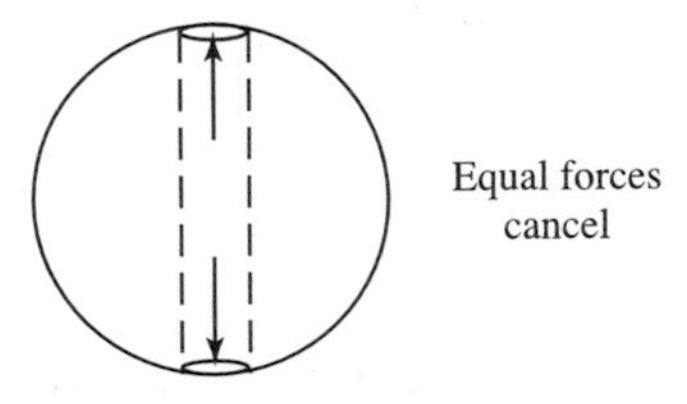

Figure 1.2. Pressure forces on opposing surfaces cancel.

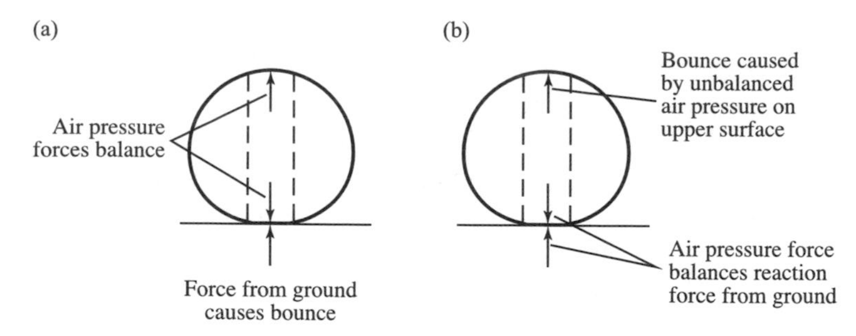

Figure 1.3. Two descriptions of the force balance during the bounce.

on the ground and, from Newton's third law, the ground exerts an equal and opposite pressure on the casing. There are two ways of viewing the resultant forces.

In the first, and more intuitive, we say that it is the upward force from the ground which first slows the ball and then accelerates it upwards, producing the bounce. In this description the air pressure force on the deformed casing is still balanced by the pressure on the opposite surface, as shown in figure 1.3(a). In the second description we say that there is no resultant force acting on the casing in contact with the ground, the excess air pressure inside the ball balancing the reaction force from the ground. The force which now causes the bounce is that of the unbalanced air pressure on that part of the casing opposite to the contact area, as illustrated in figure 1.3(b). These two descriptions are equally valid.

Because the force on the ball is proportional to the area of contact with the ground and the area of contact is itself determined by the distance of the centre of the ball from the ground, it is possible to calculate the motion of the ball. The result is illustrated in the graph of figure 1.4 which plots the height of the centre of the ball against time.

As we would expect, the calculation involves the mass and radius of the ball and the excess pressure inside it. These are precisely the quantities specified by the rules governing the ball. It is perhaps surprising that these are the only

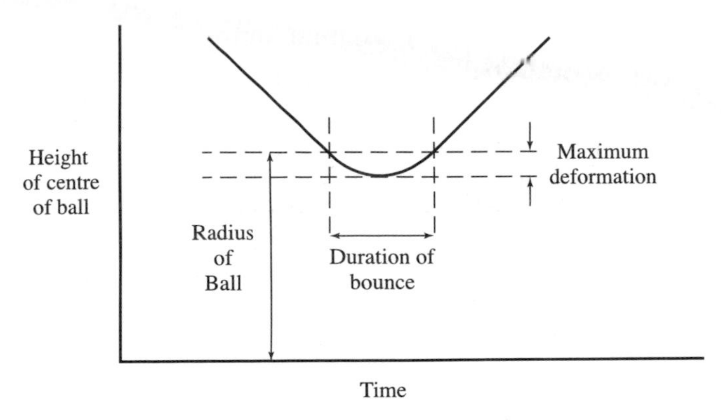

Figure 1.4. Motion of ball during bounce.

quantities involved, and that the rules determine the duration of the bounce. This turns out to be just under a hundredth of a second. The bounce time is somewhat shorter than the framing time of television pictures and in television transmissions the brief contact with the ground is often missed. Fortunately our brain fills in the gap for us.

Apart from small corrections the duration of the bounce is independent of the speed of the ball. A faster ball is more deformed but the resulting larger force means that the acceleration is higher and the two effects cancel. During the bounce the force on the ball is quite large. For a ball falling to the ground at 35 miles per hour the force rises to a quarter a ton – about 500 times the weight of the ball.

The area of casing in contact with the ground increases during the first half of the bounce. The upward force increases with the area of contact, and so the force also increases during the first half of the bounce. At the time of maximum deformation, and therefore maximum force, the ball's vertical velocity is instantaneously zero. From then on the process is reversed, the contact area decreasing and the force falling to zero as the ball loses contact with the ground.

If the ball were perfectly elastic and the ground completely rigid, the speed after a vertical bounce would be equal to that before the bounce. In reality the speed immediately after the bounce is somewhat less than that immediately before the bounce, some of the ball's energy being lost in the deformation. The lost energy appears in a very slight heating of the ball. The change in speed of the ball in the bounce is conveniently represented by a quantity called the 'coefficient of restitution'. This is the ratio, usually written e, of the speed after a vertical bounce to that before it,

$$e = \frac{\text{speed after}}{\text{speed before}}.$$

A perfectly elastic ball bouncing on a hard surface would have $e = 1$ whereas a completely limp ball which did not bounce at all would have $e = 0$. For a football on hard

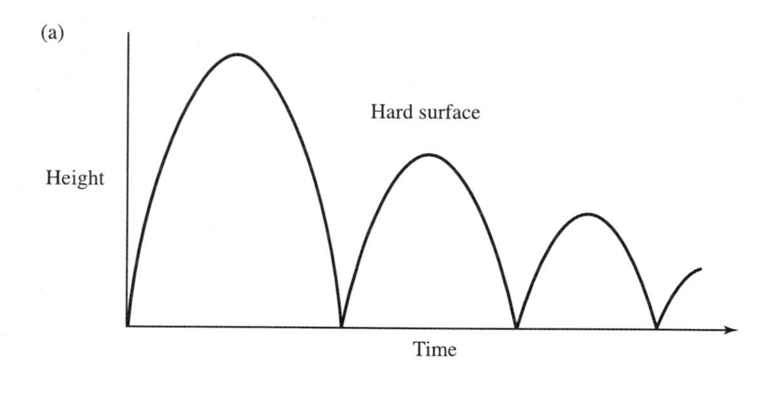

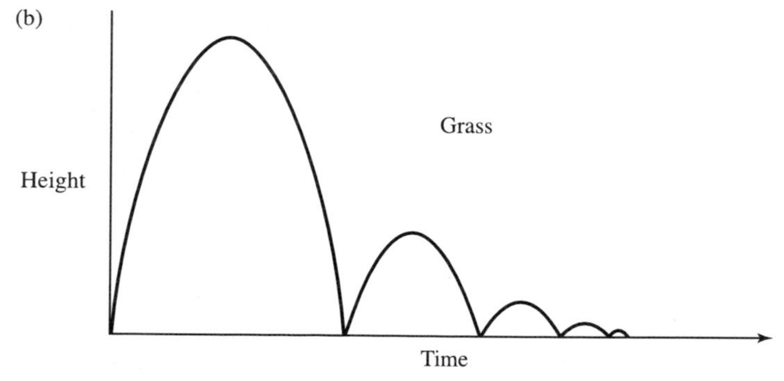

Figure 1.5. Showing how the bouncing changes with the coefficient of restitution.

ground e is typically 0.8, the speed being reduced by 20%. Grass reduces the coefficient of restitution, the bending of the blades causing further energy loss. For long grass the resulting coefficient depends on the speed of the ball as well as the length of the grass.

Figure 1.5(a) shows a sequence of bounces for a hard surface ($e = 0.8$). This illustrates the unsatisfactory nature of too bouncy a surface. Figure 1.5(b) shows the much more rapid decay of successive bounces for a ball bouncing on short grass ($e = 0.6$).

The bounce in play

The bounce described above is the simple one in which the ball falls vertically to the ground. In a game, the ball also has a horizontal motion and this introduces further aspects of the bounce. In the ideal case of a perfectly elastic ball bouncing on a perfectly smooth surface the horizontal velocity of the ball is unchanged during the bounce and the vertical velocity takes a value equal and opposite to that before the bounce, as shown in figure 1.4. The symmetry means that the angle to the ground is the same before and after. In reality the bounce is affected by the imperfect elasticity of the ball, by the friction between the ball and the ground, and by spin. Even if the ball is not spinning before the bounce, it will be spinning when it leaves the ground. We will now analyse in a simplified way the effect of these complications on the bounce.

In the case where the bounce surface is very slippy, as it would be on ice for example, the ball slides throughout the bounce and is still sliding as it leaves the ground. The motion is as shown in figure 1.6. The coefficient of restitution has been taken to be 0.8 and the resulting reduction in vertical velocity after the bounce has lowered the angle of the trajectory slightly.

In the more general case the ball slides at the start of the bounce, and the sliding produces friction between the ball and

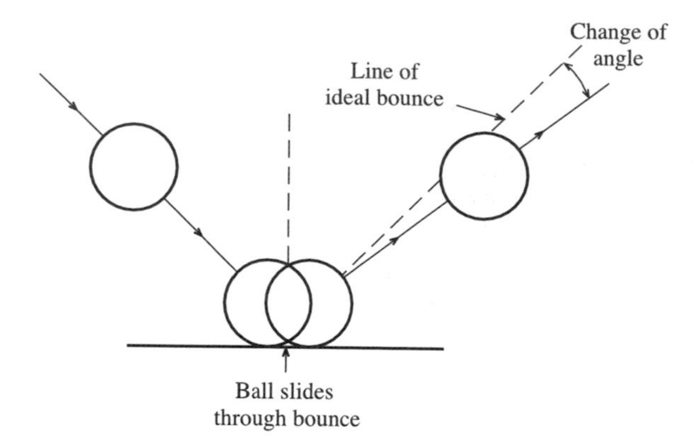

Figure 1.6. Bounce on a slippy surface.

the ground. There are then two effects. Firstly the friction causes the ball to slow, and secondly the ball starts to rotate, as illustrated in figure 1.7. The friction slows the bottom surface of the ball, and the larger forward velocity of the upper surface then gives the ball a rotation.

If the surface is sufficiently rough, friction brings the bottom surface of the ball to rest. This slows the forward motion of the ball but, of course, does not stop it. The ball then rolls about the contact with the ground as shown in figure 1.8. Since the rotation requires energy, this energy must come from the forward motion of the ball. Finally, the

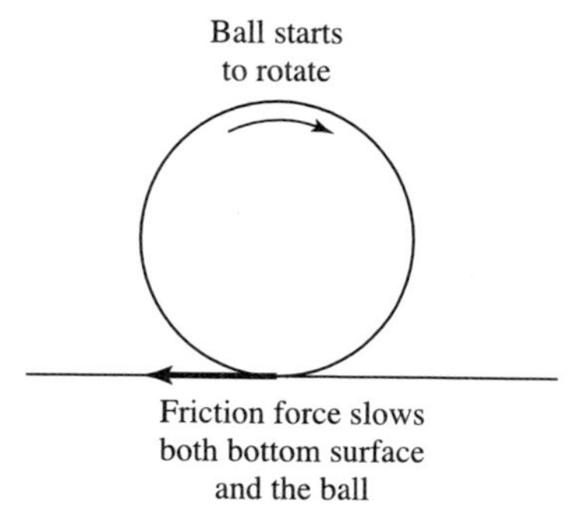

Figure 1.7. Friction slows bottom surface causing the ball to rotate.

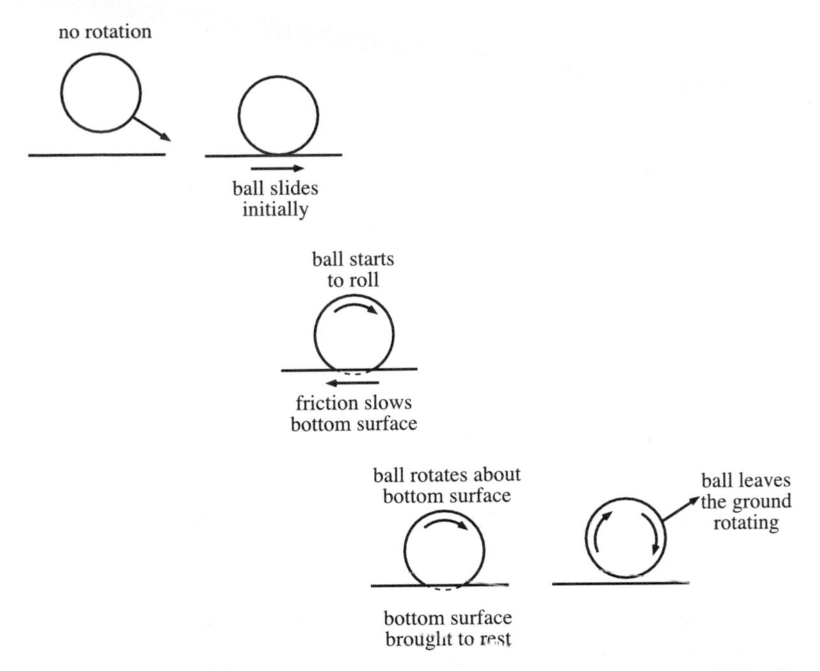

Figure 1.8. Sequence of events when the ball bounces on a surface sufficiently rough that initial sliding is replaced by rolling.

now rotating ball leaves the ground. For the case we have considered it is possible to calculate the change in the horizontal velocity resulting from the bounce. It turns out that the horizontal velocity after the bounce is three fifths of the initial horizontal velocity, the lost energy having gone into rotation and frictional heating.

Television commentators sometimes say of a ball bouncing on a slippy wet surface that it has 'speeded up' or 'picked up pace'. This is improbable. It seems likely that we have become familiar with the slowing of the ball at a bounce, as described above, and we are surprised when on a slippy surface it doesn't occur, leaving the impression of speeding up.

Whether a ball slides throughout the bounce, or starts to roll, depends partly on the state of the ground. For a given

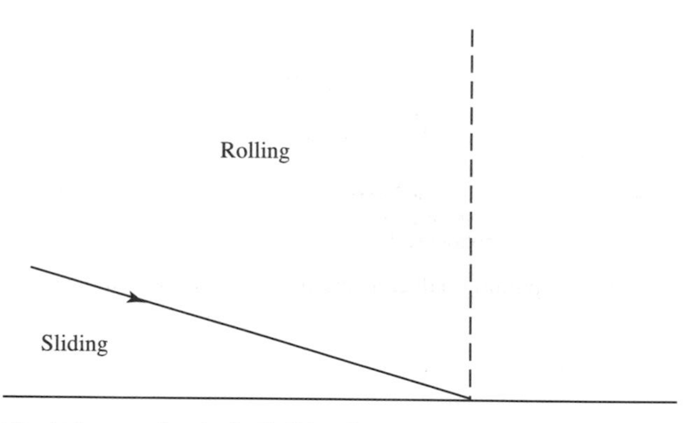

Figure 1.9. At low angles the ball slides throughout the bounce, at higher angles it rolls before it leaves the ground.

surface the most important factor is the angle of impact of the ball. For a ball to roll there must be a sufficient force on the ground and this force increases with the vertical component of the velocity. In addition, it is easier to slow the bottom surface of the ball to produce rolling if the horizontal velocity is low. Combining these two requirements, high vertical velocity and low horizontal velocity, it is seen that rolling requires a sufficiently large angle of impact. At low angles the ball slides and, depending on the nature of the ground, there is a critical angle above which the ball rolls as illustrated in figure 1.9.

With a ball that is rotating before the bounce the behaviour is more complicated, depending on the direction and magnitude of the rotation. Indeed, it is possible for a ball to actually speed up at a bounce, but this requires a rotation which is sufficiently rapid that the bottom surface of the ball is moving in the opposite direction to the motion of the ball itself as shown in figure 1.10. This is an unusual circumstance which occasionally arises with a slowly moving ball, or when the ball has been spun by hitting the underside of the crossbar.

Players can use the opposite effect of backspin on the ball to slow a flighted pass at the first bounce. The backspin slows the run of the ball and can make it easier for the receiving player to keep possession.

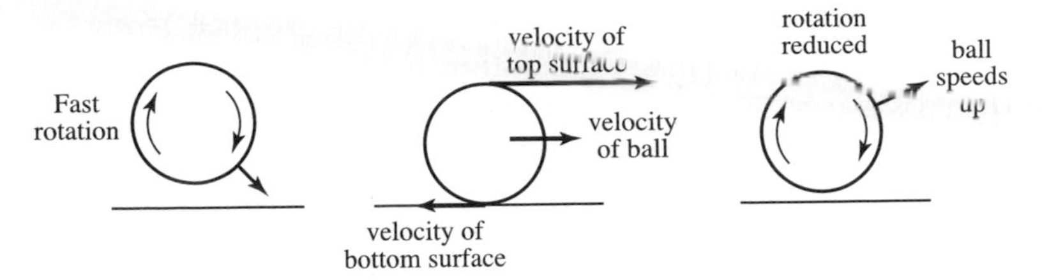

Figure 1.10. A fast spinning ball can 'speed up' during the bounce.

Bounce off the crossbar

When the ball bounces off the crossbar, the bounce is very sensitive to the location of the point of impact. The rules specify that the depth of the bar must not exceed 5 inches, and an inch difference in the point of impact has a large effect.

Figure 1.11(a) shows four different bounce positions on the underside of a circular crossbar. For the highest the top of the ball is 1 inch above the centre of the crossbar and the other positions of the ball are successively 1 inch lower. Figure 1.11(b) gives the corresponding bounce directions, taking the initial direction of the ball to be horizontal and the coefficient of restitution to be 0.7. It is seen that over the 3 inch range in heights the direction of the ball after the bounce changes by almost a right angle.

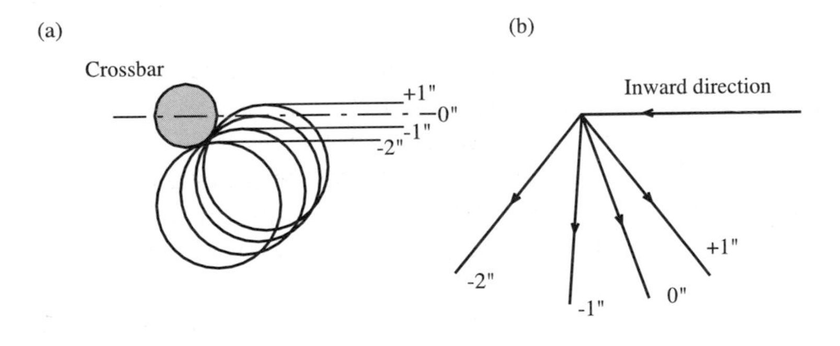

Figure 1.11. Bounce from the crossbar. (a) Positions of bounce. (b) Angles of bounce.

As with a bounce on the ground, the bounce from the crossbar induces a spin. Calculation shows that a ball striking the crossbar at 30 miles an hour can be given a spin frequency of around 10 revolutions per second. This corresponds to the lowest of the trajectories in figure 1.11. For even lower trajectories the possibility of slip between the ball and bar arises.

When the ball reaches the ground the spin leads to a change in horizontal velocity during the bounce. For example, the 30 miles per hour ball which is deflected vertically downward is calculated to hit the ground with a velocity of about 26 miles per hour and a spin of 9 rotations per second. After the bounce on the ground the ball moves away from the goal, the spin having given it a forward velocity of about 6 miles per hour.

This, of course, is reminiscent of the famous 'goal' scored by England against Germany in the 1966 World Cup Final. In that case the ball must have struck quite low on the bar, close to the third case of figure 1.11. The ball fell from the bar to the goal-line and then bounced forward, to be headed back over the bar by a German defender. Had the ball struck the bar a quarter of an inch lower it would have reached the ground fully over the line.

Chapter 2

2

The kick

The ball is kicked in a variety of ways according to the circumstances. For a slow accurate pass the ball is pushed with the flat inside face of the foot. For a hard shot the toes are dipped and the ball is struck with the hard upper part of the foot. The kick is usually aimed through the centre of the ball, but in some situations it is an advantage to impart spin to the ball. Backspin is achieved by hitting under the centre of the ball, and sidespin by moving the foot across the ball during the kick.

For a hard kick, such as a penalty or goal kick, there are two basic elements to the mechanics. The first is the swinging of the leg to accelerate the foot, and the second is the brief interaction of the foot with the ball. Roughly, the motion of the foot takes a tenth of a second and the impact lasts for a hundredth of a second.

For the fastest kicks the foot has to be given the maximum speed in order to transfer a high momentum to the ball. To achieve this the knee is bent as the foot is taken back. This allows the foot to be accelerated through a long trajectory, producing a high final speed. The muscles accelerate the thigh, pivoting it about the hip, and accelerate even faster the calf and the foot. As the foot approaches impact with the ball the leg straightens, and at impact the foot is locked firmly with the leg. This sequence is illustrated in figure 2.1.

If the interaction of the foot with the ball were perfectly elastic, with no frictional energy losses, the speed given to

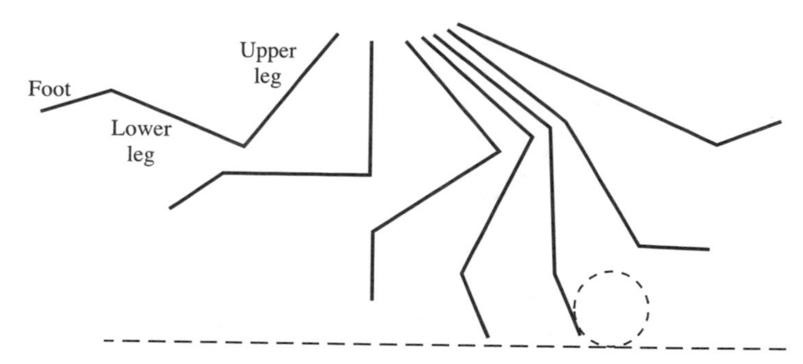

Figure 2.1. In a fast kick the upper leg is driven forward and the lower leg whips through for the foot to transfer maximum momentum to the ball.

the ball would follow simply from two conservation laws. The first is the conservation of energy and the second is the conservation of angular momentum. These laws determine the fall in speed of the foot during the impact, and the resulting speed of the ball. If, further, the mass of the ball is taken to be negligible compared with the effective mass of the leg, the speed of the foot would be unchanged on impact. In this idealised case, the ball would then 'bounce' off the foot and take a speed equal to twice that of the foot.

In reality the leg and the foot are slowed on impact and this reduces the speed of the ball. Frictional losses due to the deformation of the ball cause a further reduction in speed. This reduction can be allowed for by a coefficient of restitution in a similar way to that for a bounce. When these effects are taken into account it turns out that at the start of the impact the foot is moving at a speed about three-quarters of the velocity imparted to the ball. This means that for a hard kick the foot would be travelling at more than 50 miles per hour.

Mechanics

It was seen in figure 2.1 that in a hard kick the thigh is forced forward and the calf and the foot are first pulled forward and

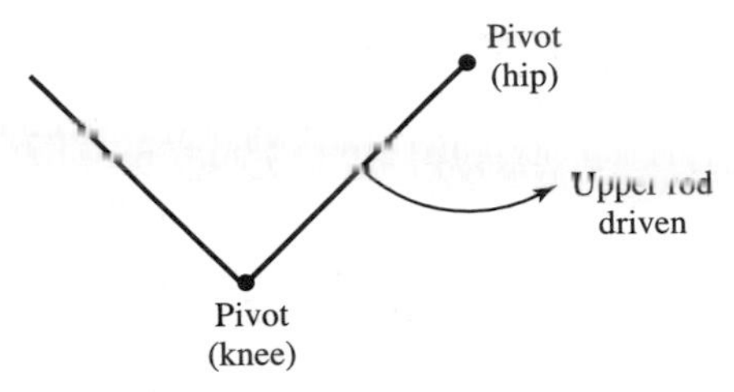

Figure 2.2. Model in which the upper and lower parts of the leg are represented by two pivoting rods, the upper of which is driven around the (hip) pivot.

then swing through to strike the ball. The mechanics of the process can be illustrated by a simple model in which the upper and lower parts of the leg are represented by rods and the hip and knee are represented by pivots, as illustrated in figure 2.2. Let us take the upper rod to be pulled through with a constant speed and ask how the lower rod, representing the lower leg, moves. Figure 2.3 shows what happens. Initially the lower rod is pulled by the lower pivot and moves around with almost the same speed as the pivot. However, the centrifugal force on the lower rod 'throws' it outward, making it rotate about the lower pivot and increasing its speed as it

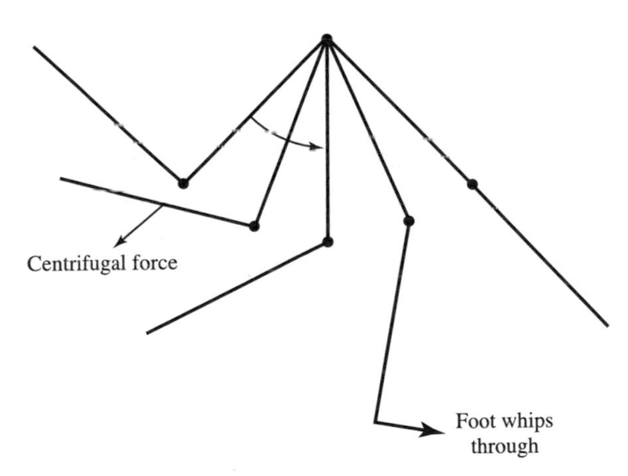

Figure 2.3. The lower rod is pulled around by the upper rod and is thrown outward by the centrifugal force, accelerating the foot of the rod.

does so. As the upper rod moves round, the lower rod 'whips' around at an increasing rate and in the final stage illustrated the two rods form a straight line. The whipping action gives the foot of the lower rod a speed about three times that of the lower (knee) pivot.

This model represents quite well the mechanics of the kick. The motion illustrated by the model is familiar as that of the flail used in the primitive threshing of grain, and is also similar to that of the golf swing. When applied to golf the upper rod represents the arms and the lower rod represents the club.

Since students of elementary physics are sometimes confused by the term centrifugal force used above, perhaps some comment is in order. When a stone is whirled around at the end of a string it is perfectly proper to say that the force from the string prevents the stone from moving in a straight line by providing an inward acceleration. But it is equally correct to say that from the point of view of the stone the inward force from the string balances the outward centrifugal force. This description is more intuitive because we have experienced the centifugal force ourselves, for example when in a car which makes a sharp turn.

Forces on the foot

During the kick there are three forces on the foot, as illustrated in figure 2.4. Firstly, there is the force transmitted from the leg to accelerate the foot towards the ball. Secondly, and particularly for a hard kick, there is the centrifugal force as the foot swings through an arc. The third force is the reaction from the ball which decelerates the foot during impact.

To see the magnitude of these forces we take an example where the foot is accelerated to 50 miles per hour over a distance of 3 feet. In this case the force on the foot due to acceleration is 30 times its weight and the centrifugal force reaches a somewhat greater value. On impact with the ball the foot's speed is only reduced by a fraction, but this

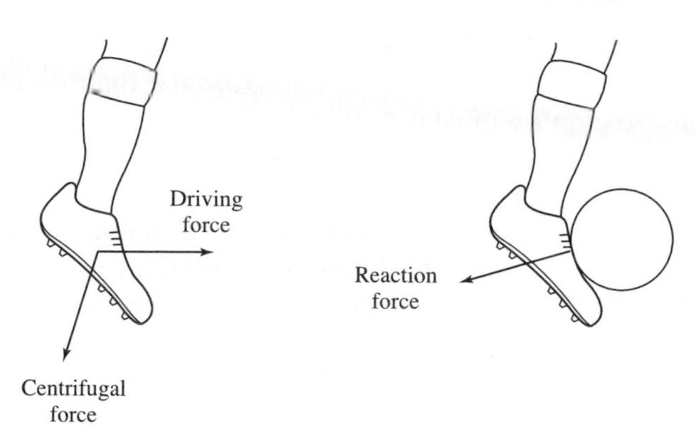

Figure 2.4. The three forces on the foot during a kick.

occurs on a shorter timescale than that for its acceleration and the resulting deceleration force on the foot during impact is about twice the force it experiences during its acceleration.

Power

The scientific unit of power is the watt, familiar from its use with electrical equipment. It is, however, common in English speaking countries to measure mechanical power in terms of horse-power, the relationship being 1 horse-power = 750 watts. The name arose when steam engines replaced horses. It was clearly useful to know the power of an engine in terms of the more familiar power of horses. As would be expected, human beings are capable of sustaining only a fraction of a horse-power. A top athlete can produce a steady power approaching half a horse-power.

The muscles derive their power from burning glucose stored in the muscle, using oxygen carried from the lungs in the bloodstream. The sustainable power is limited by the rate of oxygen intake to the lungs, but short bursts of power can use a limited supply of oxygen which is immediately available in the muscle. This allows substantial transient powers to be achieved. What is the power developed in a kick?

Both the foot and the leg are accelerated, and the power generated by the muscles is used to produce their combined kinetic energy. For a fast kick the required energy is developed in about a tenth of a second, and the power is calculated by dividing the kinetic energy by this time. It turns out that about 10 horse-power is typically developed in such a kick.

The curled kick

To produce a curved flight of the ball, as illustrated in figure 2.5, it is necessary to impart spin to the ball during the kick. The spin alters the airflow over the ball and the resulting asymmetry produces a sideways force which gives the ball its curved trajectory. We shall look at the reason for this in chapter 4. Viewed from above, a clockwise spin curls the ball to the right, and an anticlockwise spin to the left.

Figure 2.6(a) shows how the foot applies the necessary force by an oblique impact. This sends the ball away spinning and moving at an angle to the direction of the target. The ball then curls around to the target as shown in figure 2.6(b). The amount of bend depends upon the spin rate given to the ball,

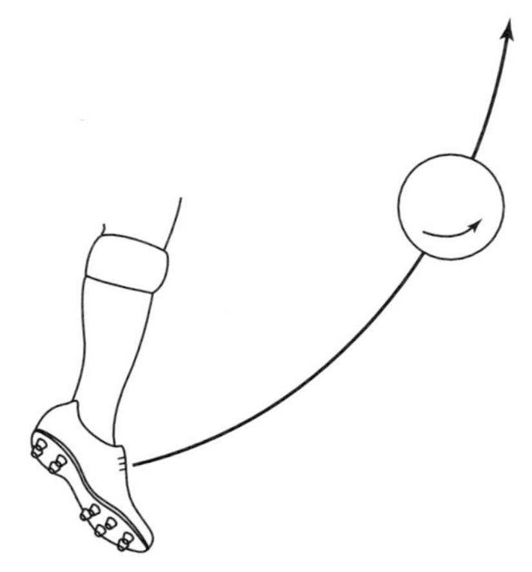

Figure 2.5. Curved flight of spun ball.

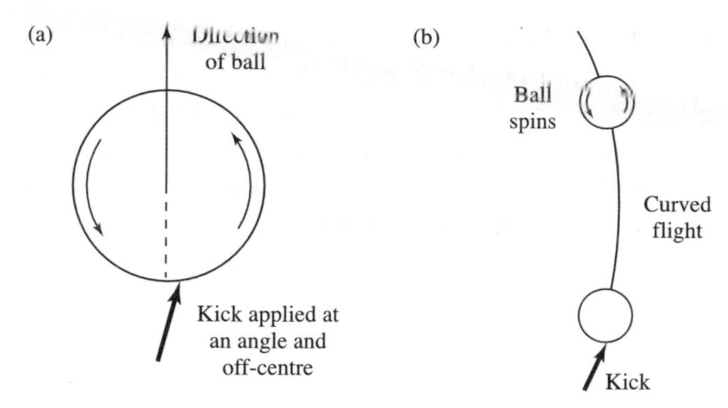

Figure 2.6. To produce a curved flight the ball is struck at an angle to provide the necessary spin.

and the skill lies in achieving the required rotation together with accuracy of direction. An analysis of the mechanics of the kick is given in chapter 10.

Only a small part of the energy transferred to the ball is required to produce a significant spin. If the energy put into the spin in a 50 mile per hour kick is 1% of the directed energy, the ball would spin at 4 revolutions per second.

Accuracy

The directional accuracy of a kick is simply measured by the angle between the direction of the kick and the desired direction. However, it is easier to picture the effect of any error by thinking of a ball kicked at a target 12 yards away. This is essentially the distance faced by a penalty taker. Figure 2.7 gives a graph of the distance by which the target would be missed for a range of errors in the angle of the kick.

There are two sources of inaccuracy in the kick, both arising from the error in the force applied by the foot. The first contribution comes from the error in the direction of the applied force and the second from misplacement of the

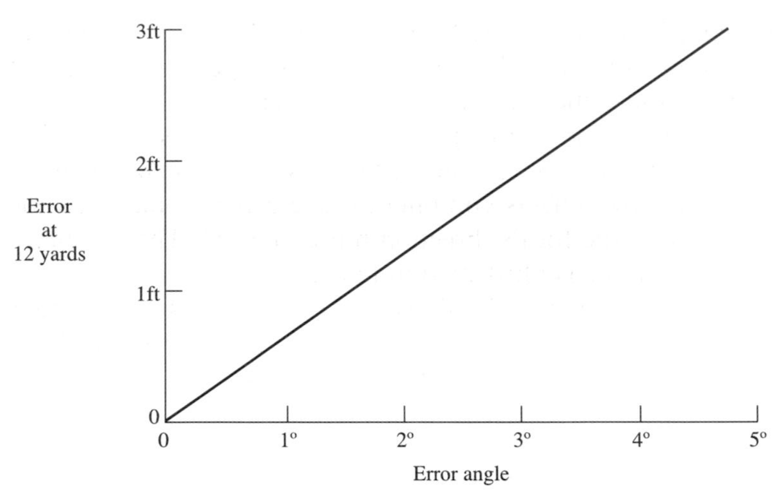

Figure 2.7. Error at a distance of 12 yards resulting from a given error in the direction of the kick.

force. These two components are illustrated separately in figure 2.8.

It is seen from figure 2.7 that placing the ball within one yard at a distance of 12 yards requires an accuracy of angle of direction of the ball of about 5°. The required accuracy of direction for the foot itself is less for two reasons. Firstly,

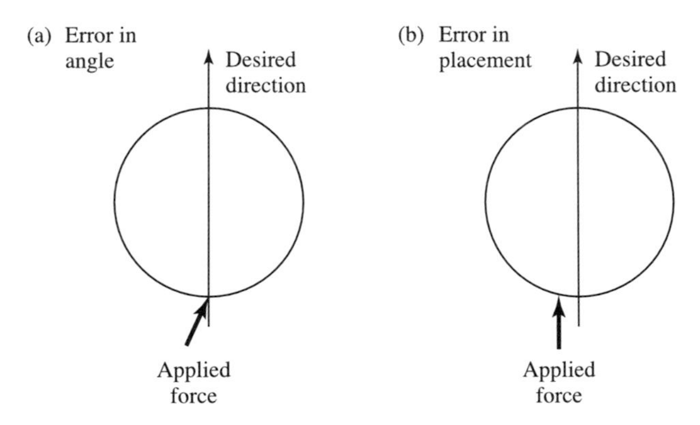

Figure 2.8. The kick can have errors in both direction and placement on the ball. In (a) and (b) these are shown separately.

the ball bounces off the foot with a forward velocity higher than that of the foot by a factor depending on the coefficient of restitution and, secondly, part of the energy supplied by the sideways error force goes into rotation of the ball rather than sideways velocity. For a 5% accuracy of the ball's direction these two effects combine to give a requirement on the accuracy of the foot's direction more like 15°. The geometry of this example is illustrated in figure 2.9.

The accuracy of the slower side-foot kick is much better than that of the fast kick struck with the top of the foot. Because of the flatness of the side of the foot the error from

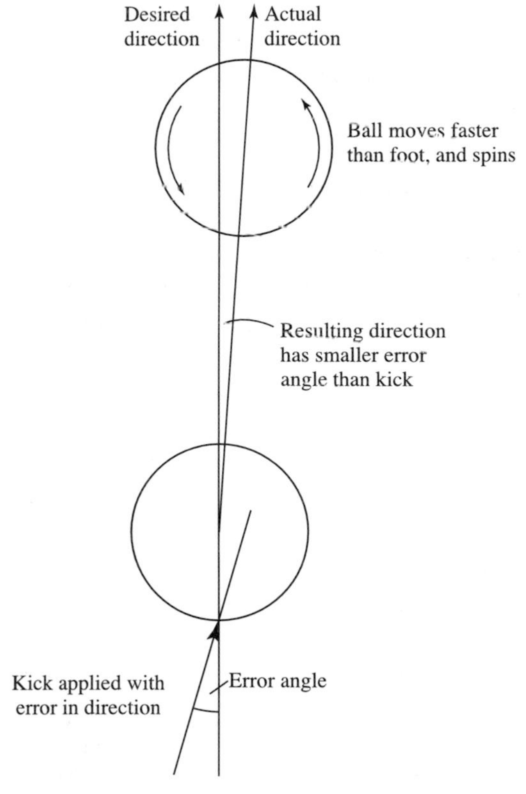

Figure 2.9. When there is an error in the direction of the applied kick the error in the direction of the ball is much less.

placement of the foot on the ball is virtually eliminated, leaving only the error arising from the direction of the foot. This makes the side-foot kick the preferred choice when accuracy is more important than speed.

How fast?

The fastest kicks are normally unhindered drives at goal, the obvious case being that of a penalty-kick struck with maximum force. To take an actual case we can look at the penalty shoot-out between England and Germany in the 1996 European Championships. Twelve penalty-kicks were taken and the average speed of the shots was about 70 miles per hour. The fastest kick was the last one, by Möller, with a speed of about 80 miles per hour. Goal-kicks usually produce a somewhat lower speed, probably because of the need to achieve range as well as speed.

It is possible to obtain a higher speed if the ball is moving towards the foot at the time of impact. The speed of the foot relative to the ball is increased by the speed of the incoming ball and consequently the ball 'bounces' off the foot with a higher speed. When allowance is made for the unavoidable frictional losses and the loss of momentum of the foot, the

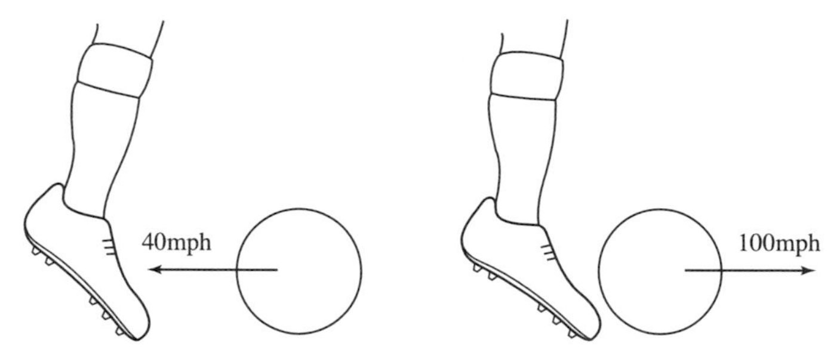

Figure 2.10. A kick produces a higher ball speed when the ball is initially moving toward the foot. In this example the kick is such that it would give a stationary ball a speed of 80 miles per hour.

increment in the speed of the ball leaving the kick is about half the incoming speed of the ball. Taking a kick which would give a stationary ball a speed of 80 miles per hour we see that a well-struck kick with the ball moving toward the player at 40 miles per hour, which returns the ball in the direction from which it came, could reach $80 + \frac{1}{2}40 = 100$ miles per hour, as illustrated in figure 2.10.

Chapter 3

3

Throwing, heading, punching, catching, receiving, trapping

Acceleration, *g*, and forces

The subjects of this chapter are all concerned with acceleration or deceleration of the ball. In order to give some intuitive feel for the accelerations and forces involved the accelerations will be expressed in terms of the acceleration due to gravity, which is written as g, and forces will be described by the force of an equivalent weight. Because most British people think of speeds in terms of miles per hour and weight in terms of pounds these units will be used. In scientific work the basic units are the metre, kilogram and second and in the final, theoretical, chapter we shall change to these units.

Objects falling freely under gravity have an acceleration of 22 miles per hour per second (9.8 metres per second per second), so in each second the vertical velocity increases by 22 miles per hour. Thus an acceleration of 220 miles per hour per second is $10g$.

Forces will be given in pounds. For example a force of 140 pounds is equal to the gravitational force of 140 pounds weight (10 stone). The gravitational force on an object produces an acceleration g and, correspondingly, an acceleration, g, of the object requires a force equal to its weight. Similarly, to accelerate an object by $10g$, for example, requires a force equal to 10 times its weight.

Conversion table

1 yard	= 0.91 metre
1 mile/hour	= 1.47 feet/second
	= 0.45 metre/second
1 pound	= 0.45 kilogram

The throw-in

Usually the throw-in is used to pass the ball directly to a well-placed colleague. The distance thrown is generally not great and the required accuracy is easily achieved by any player. A more difficult challenge arises when the ball is to be thrown well into the penalty area to put pressure on the opponent's goal. To reach the goal-area calls for a throw approaching 30 yards, and long throws of this type often become a speciality of players with the necessary skill.

A short throw of, say, 10 yards needs a throw speed of around 20 miles per hour. Taking a hand movement of 1 foot the required force is typically 10–15 pounds.

A throw to the centre of the pitch, as illustrated in figure 3.1, requires a throw of almost 40 yards. In the absence of air resistance this challenging throw would require the ball to be thrown with a speed of 40 miles per hour. The effect of air drag increases the required speed to about 45 miles per

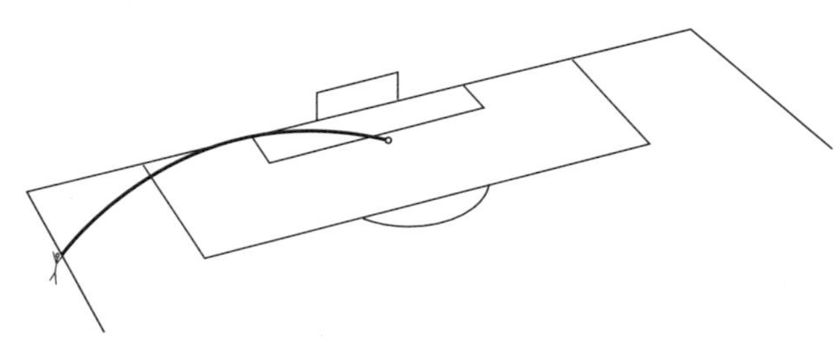

Figure 3.1. Throw to centre of the pitch.

hour. To give the ball such a high speed the thrower must apply a large force over as long a path as possible. Although a short run up to the throwing position is helpful, both feet must be in contact with the ground during the throw. This limits the distance the arms can move. The back is initially arched with the ball behind the head, and the muscles of the body and arms are then used to push the ball forward and upward. For a long throw the ball remains in contact with the hands over a distance of about 2 feet. Taking this figure the average acceleration of the ball needed to reach 45 miles per hour is 34*g*. Since the ball weighs approximately a pound this means that the average force on the ball must be about 34 pounds; the maximum force will of course be somewhat larger.

The record for the longest throw was achieved by the American college player Michael Lochnor, who threw the ball 52.7 yards in 1998. The record was previously held by David Challinor of Tranmere Rovers who reached 50.7 yards, and this throw remains the British record.

Goalkeeper's throw

Goalkeepers often trust their throw rather than their kick. The ball can be quite accurately rolled or thrown to a nearby colleague. Sometimes the goalkeeper chooses to hurl the ball toward the half-way line rather than kick it, and an impressive range can be obtained in this way. Despite the use of only one arm these throws can carry farther than a throw-in. This is partly because of the longer contact with the ball during the throw, allowing the force to be applied for more time, and partly because of the greater use of the body muscles. The greater ease of obtaining the optimum angle of throw for a long range is probably another factor. For a long throw the hand remains in contact with the ball for about 6 feet, and the contact time for the throw is typically several times as long as for a throw-in.

Heading

A well-headed ball is struck with the upper part of the forehead and the ball essentially bounces from the head. The types of header are characterised by the way in which momentum is transferred between the head and the ball.

When a defender heads away a long ball his neck is braced and the bounce of the ball from his head transfers momentum to his body. Another situation in which momentum is taken by the body is in the diving header. In this case the whole body is launched at the ball and it is the speed of the body which determines the resulting motion of the ball.

In more vigorous headers the muscles are used to thrust the head at the ball. This type of header is commonly used by strikers to propel a cross from the side of the pitch toward the goal. When the head strikes the ball, momentum is transferred to the ball and the head is slowed. Because the head weighs several times as much as the ball and because it is anchored at the neck the change in speed of the head through the impact is typically less than 10% of the speed given to the ball. In heading the ball the movement of the head is restricted to a few inches, and the velocity given to the ball is much less than that possible for a kick.

Sometimes the head is struck by an unseen ball, or before the player can prepare himself. It is then possible for all the ball's loss of momentum to be transferred to the head. In a severe case of a 50 mile per hour ball, the head could be moved an inch in a hundredth of a second, the force on the head corresponding to an acceleration of 50g. Accelerations larger than this can lead to unconsciousness.

The punch

Wherever possible, goalkeepers aim to take charge of a ball close to goal by catching it. There are two circumstances where this is not possible. Firstly there is the ball which is flighted into a group of players near the goal and goalkeeper doesn't have sufficient access to the ball to be confident of catching it. If he can he will then punch the ball as far away from the goal as possible. The punch is less powerful than the kick and the distance of movement of the fist is limited to about a foot. However, the ball bounces off the fist, taking a higher speed than the fist speed. Typically a range of about 20 yards is obtained, corresponding to a fist speed of about 20 miles per hour.

The second situation where a punch is called for is where a shot is too far out of the goalkeeper's reach for a catch to be safely made and a punch is the best response. When the punch follows a dive by the goalkeeper, considerable accuracy is called for because of the brief time that a punch is possible. For example, a ball moving at 50 miles per hour passes through its own diameter in one hundredth of a second.

While the punch is usually the prerogative of the goalkeeper, it is also possible to score a goal with a punch. Figure 3.2 shows a well-known instance of this.

The catch

Goalkeepers make two kinds of catch. The simpler kind is the catch to the body. In this case most of the momentum of the ball is transferred to the body. Because of the comparatively large mass of the body the ball is brought to rest in a short distance. The goalkeeper then has to trap the ball with his hands to prevent it bouncing away.

In the other type of catch the ball is taken entirely with the hands. With regard to the mechanics, this catch is the inverse of a throw. The ball is received by the hands with its incoming speed and is then decelerated to rest. During the deceleration the momentum of the ball is transferred to the hands and arms through the force on the hands. The skill in this catch is to move the hands with the ball while it is brought to rest. Too small a hand movement creates a too rapid deceleration

Figure 3.2. Maradona bending the rules. (© *Popperfoto/Bob Thomas Sports Photography*.)

of the ball and the resulting large force makes the ball difficult to hold. The movement of the hands during the catch is nevertheless usually quite small, typically a few inches.

Taking as an example a shot with the ball moving at 50 miles per hour, and the goalkeeper's hands moving back 6 inches during the catch, the average deceleration of the ball is 170*g*, so the transient force on the hands is 170 pounds, which is roughly the weight of the goalkeeper. The catch is completed in just over a hundredth of a second.

Receiving

When a pass is received by a player the ball must be brought under control, and in tight situations this must be done without giving opponents a chance to seize the ball. The basic problem with receiving arises when the ball comes to the player at speed. If the ball is simply blocked by the foot, it bounces away with a possible loss of possession. The ball is controlled by arranging that the foot is moving in the same direction as the ball at the time of impact. The mechanics are quite straightforward – essentially the same as for a bounce, but with a moving surface. Thus, allowing for the coefficient of restitution, the speed of the foot can be chosen to be such that the ball is stationary after the bounce. It turns out that the rule is that the foot must be moving at a speed equal to the speed of the ball multiplied by $e/(1+e)$ where is the coefficient of restitution. If, say, the ball is moving at a speed of 25 miles per hour and the coefficient of restitution is $\frac{2}{3}$, then the foot must be moving back at a speed of 10 miles per hour. This ideal case, where the ball is brought to rest, is illustrated in figure 3.3.

To receive a fast ball successfully it is not only necessary to achieve the correct speed of the foot, but also requires good timing. A ball travelling at 30 miles per hour moves a distance equal to its own diameter in about a sixtieth of a second, and this gives an idea of the difficulty involved. The player's reaction time is more than ten times longer than this, showing that the art lies in the anticipation.

Trapping

Trapping the ball under the foot presents a similar challenge to that of receiving a fast pass in that the time available is very brief. A particular need to trap the ball arises when it reaches the player coming downwards at a high angle. To prevent the ball bouncing away the foot is placed on top of it at the moment of the bounce. Easier said than done.

As the ball approaches, the foot must be clear of it so that the ball can reach the ground. Then, when the ball reaches the ground the foot must be instantly placed over it, trapping the

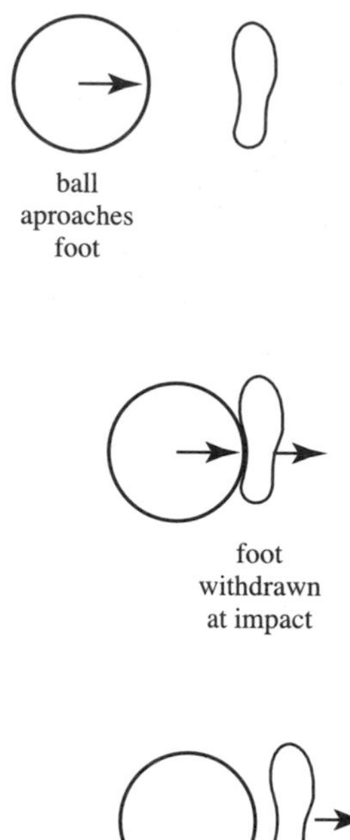

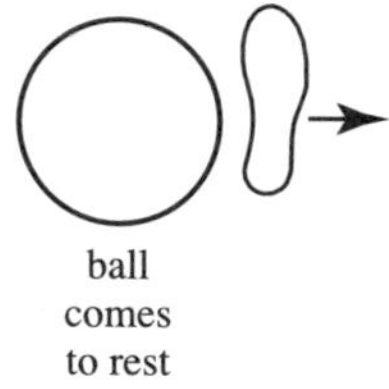

Figure 3.3. Controlling the ball.

ball between the foot and the ground. This is sometimes done with great precision. The 'window' of time within which trapping is possible is determined by the requirement that the foot is placed over the ball in the time it takes for the ball to reach the ground and bounce back up to the foot, as illustrated in figure 3.4.

We can obtain an estimate of the time available by taking the time for the top of the ball to move downwards from the level of the foot and then to move upwards to that level again. The upward velocity will be reduced by the coefficient of restitution but for an approximate answer this effect is neglected. If the vertical distance between the ball and the foot at the time of bounce is, say, 3 inches then taking a

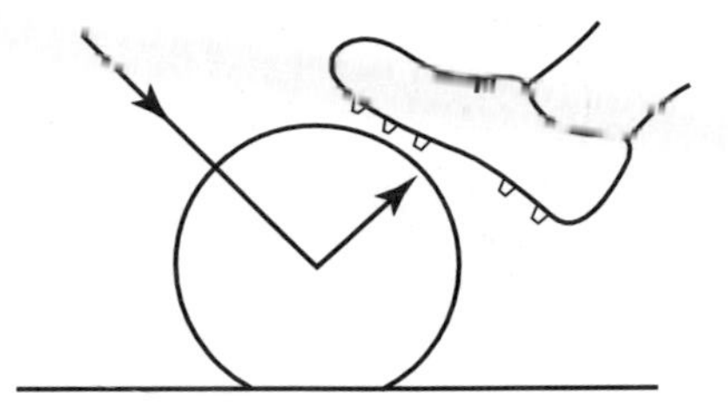

Figure 3.4. Trapping the ball requires a well timed placement of the foot.

hundredth of a second for the duration of the bounce, a ball travelling at 30 miles per hour will allow about a fiftieth of a second to move the foot into place. As with receiving a fast pass, anticipation is the essential element.

Chapter 4

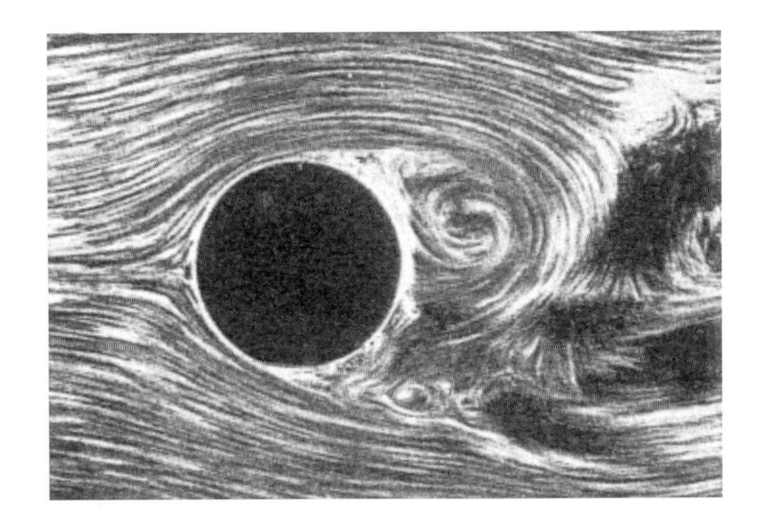

4

The ball in flight

In professional baseball and cricket, spinning the ball to produce a curved flight and deceive the batsman is a key part of the game. Footballers must have known from the early days of organised football in the nineteenth century that their ball can be made to move in a similar way. But it was the Brazilians who showed the real potential of the 'banana' shot. Television viewers watched in amazement as curled free kicks ignored the defensive wall and fooled the goalkeeper. The wonderful goals scored by Roberto Rivelino in the 1974 World Cup and by Roberto Carlos in the Tournoi de France in 1997 have become legends. This technique is now widespread, and we often anticipate its use in free kicks taken by those who have mastered the art.

We shall later look at the explanation of how a spinning ball interacts with the air to produce a curved flight, but we first look at the long range kick. What is surprising is that understanding the ordinary long range kick involves a very complicated story. Long range kicks require a high speed, and at high speed the drag on the ball due to the air becomes very important. If there were no air drag, strong goal-kicks would fly out at the far end of the pitch as illustrated in figure 4.1. In exploring the nature of air drag we shall uncover the unexpectedly complex mechanisms involved. However, we

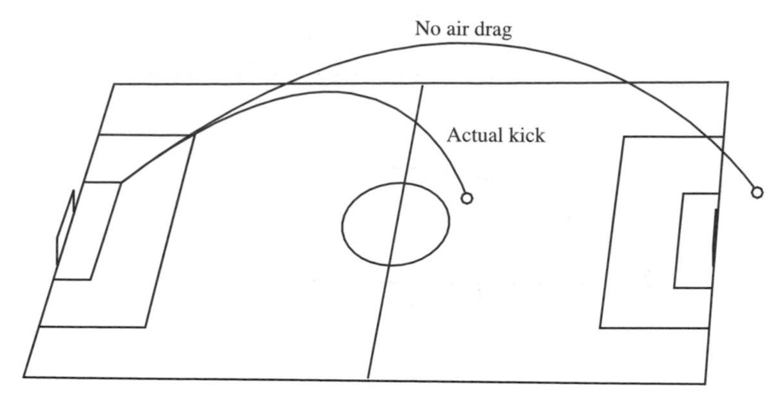

Figure 4.1. Flight of a goal-kick compared with that which would occur without air drag.

best start by looking at the idealised case of the flight of the ball without air drag.

Flight without drag

It was in the seventeenth century that the Italian astronomer and physicist Galileo discovered the shape of the curved path travelled by projectiles. He recognised that the motion could be regarded as having two parts. From his experiments he discovered that the vertical motion of a freely falling object has a constant acceleration and that the horizontal motion has a constant velocity. When he put these two parts together, and calculated the shape of the projectile's path, he found it to be a parabola.

We would now say that the vertical acceleration is due to the earth's gravity, and call the acceleration g. Everyone realised, of course, that Galileo's result only applies when the effect of the air is unimportant. It was obvious, for instance, that a feather does not follow a parabola.

When the air drag is negligible, as it is for short kicks, a football will have a parabolic path. Figure 4.2 shows the parabolas traced by balls kicked at three different angles,

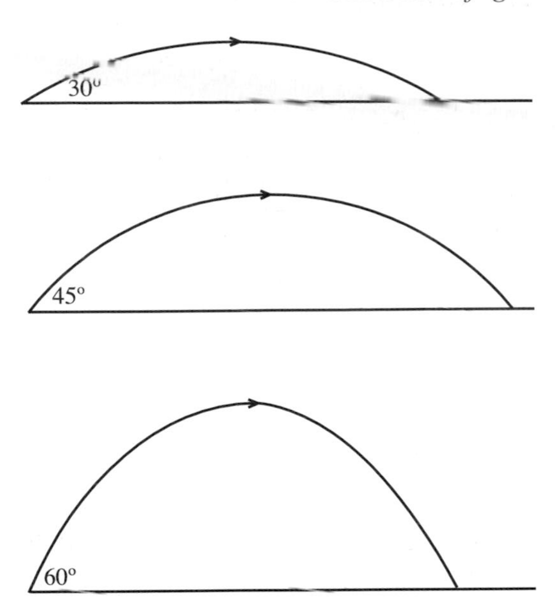

Figure 4.2. Neglecting air drag the ball flies in a parabola, the shape depending on the angle of the kick. The figure shows the paths of three balls kicked with the same initial speed but at different angles.

but with the same initial speed. The distance travelled by the ball before returning to the ground depends only on the angle and speed with which the ball leaves the foot. For a given speed the maximum range is obtained for a kick at 45°, as illustrated in the figure. The range for 30° and 60° kicks is 13% less.

To better understand this, we look at the velocity of the ball in terms of its vertical and horizontal parts. The distance the ball travels before returning to the ground is calculated by multiplying its horizontal velocity by the time it spends in the air. If the ball is kicked at an angle higher than 45°, its time in the air is increased, but this is not sufficient to compensate for the reduction in horizontal velocity, and the range is reduced. Similarly, at angles below 45° the increased horizontal velocity doesn't compensate for the reduction of the time in the air. In the extreme cases this becomes quite obvious. For a ball

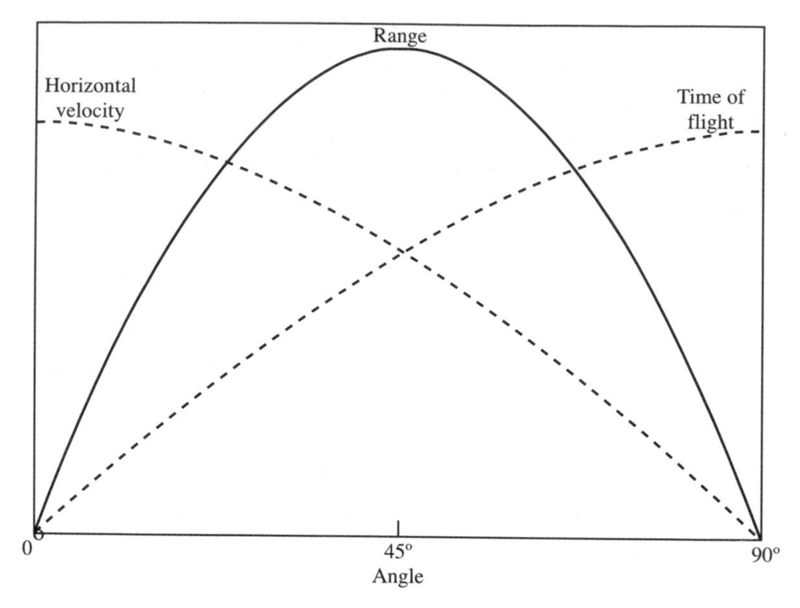

Figure 4.3. For parabolic paths the range is given by multiplying the constant horizontal velocity by the time of flight. For balls kicked with the same speed, both of these depend on the angle of the kick. As the angle of the kick is increased, the horizontal part of the velocity falls and the time of flight increases, giving a maximum at 45°.

kicked vertically the range is zero, and a ball kicked horizontally doesn't leave the ground.

These effects are brought out more fully in figure 4.3, which shows the horizontal velocity and the time of the flight for all angles. When multiplied together they give the range shown, with its maximum at 45°.

The time the ball takes to complete its flight can also be calculated. This time depends only on the vertical part of the ball's initial velocity, and the time in seconds is approximately one tenth of the initial vertical velocity measured in miles per hour. A ball kicked with an initial vertical component of velocity of 20 miles per hour would therefore be in the air for 2 seconds.

For slowly moving balls the air drag is quite small and for speeds less than 30 miles per hour the effect of air drag is not important. However, for long range kicks, such as goal-kicks, calculations ignoring the effect of the air give seriously incorrect predictions. To understand how the air affects the ball we need to look at the airflow over the ball.

The airflow

Figure 4.4 gives an idealised picture of the airflow around a ball. The airflow is shown from the 'point of view' of the ball – the ball being taken as stationary with the air flowing over it. This is a much easier way of looking at the behaviour than trying to picture the airflow around a moving ball.

The lines of flow are called streamlines. Each small piece of air follows a streamline as it flows past the ball. The air between two streamlines remains between those streamlines throughout its motion. What the figure actually shows is a cross-section through the centre of the ball. Considered in three dimensions the stream lines can be thought of as making up a 'stream surface', enclosing the ball, as shown in figure 4.5. The air arrives in a uniform flow. It is then

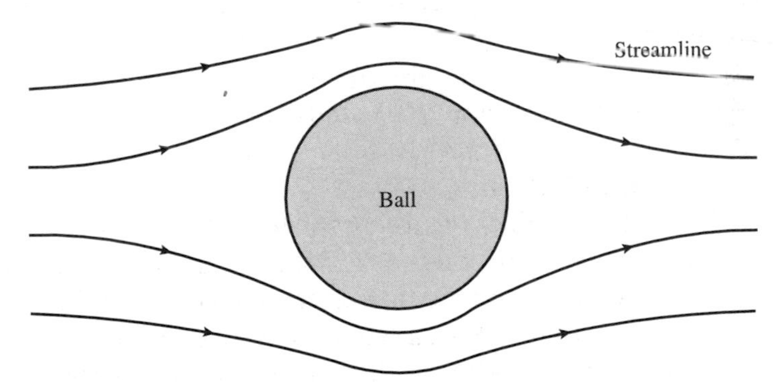

Figure 4.4. Cross-section of the airflow over the ball, the flow following the streamlines.

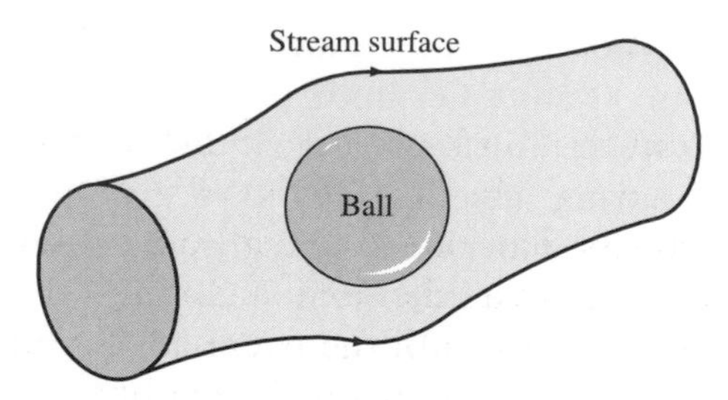

Figure 4.5. Three dimensional drawing of a stream surface, showing how the air flows around the ball.

pushed aside to flow around the ball and, in this simplified picture, returns at the back of the ball to produce a uniform flow downstream from the ball.

The surprising thing is that the simple flow described above produces no drag on the ball, a result first appreciated by the French mathematician d'Alembert in the eighteenth century. In simple terms this can be understood from the fact that the downstream flow is identical to the upstream flow, no momentum having been transferred from the air to the ball.

To understand what really happens we need to take account of the viscosity of the air. Viscosity is more easily recognised in liquids, but its effect on air can be observed, for example, when it slows the air driven from a fan, and ultimately brings it to rest.

The simplest model of viscous flow over a sphere is that given by the Irish physicist Stokes in the nineteenth century. Many physics students will have verified 'Stokes's law' for the viscous drag on a sphere, by dropping small spheres through a column of oil or glycerine. A crucial, and correct, assumption of this model is that the fluid, in our case the air, is held stationary at the surface of the sphere, so that the flow velocity at the surface is zero. The difference in velocity which then naturally arises between the slowed flow close to the ball and the faster flow further away gives rise to a viscous force, which is felt by the ball as a drag.

The boundary layer

Now it turns out that Stokes's viscous model will not explain the drag on a football. In fact the model is only valid for ball velocities much less than one mile per hour. Not much use to us. The essential step to a fuller understanding the flow around solid bodies had to wait until the twentieth century when the German physicist Prandtl explained what happens.

Imagine taking a ball initially at rest, and moving it with a gradually increasing velocity. At the beginning, the region around the ball which is affected by viscosity is large – comparable with the size of the ball itself. As the velocity is increased the viscous region contracts towards the ball, finally becoming a narrow layer around the surface. This is called the boundary layer. The drag on the ball is determined by the behaviour of this layer, and outside the layer viscosity can be neglected. With a football the boundary layer is typically a few millimetres thick, becoming narrower at high speed.

The boundary layer doesn't persist around to the back of the ball. Before the flow in the boundary layer completes its course it separates from the surface as shown in figure 4.6. Behind the separation point the flow forms a turbulent

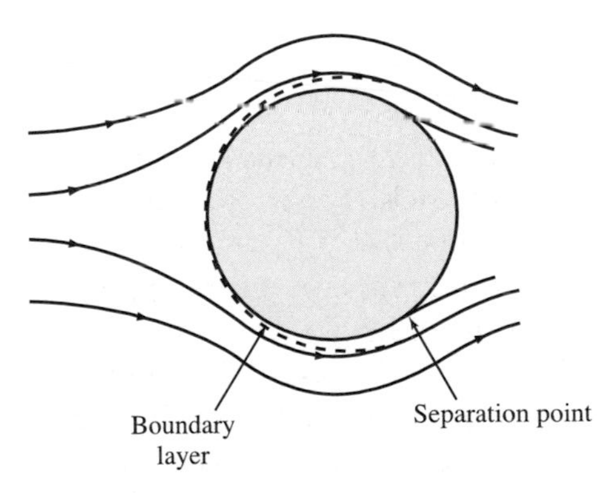

Figure 4.6. The boundary layer is a narrow region around the surface of the ball in which the effect of viscosity is concentrated. Viscosity slows the airflow causing it to separate from the ball.

wake. In this process the air in the wake has been slowed, and it is the reaction to this slowing which is the source of the air drag on the ball. In order to understand how this separation happens we must see how the velocity of the air changes as it flows around the ball, and how these changes are related to the variation of the pressure of the air. This leads us to the effect explained by the Swiss mathematician Bernoulli, and named after him.

The Bernoulli effect

Figure 4.7 shows streamlines for an idealised flow. If we look at the streamlines around the ball we see that they crowd together as the air flows around the side of the ball. For the air to pass through the reduced width of the flow channel it

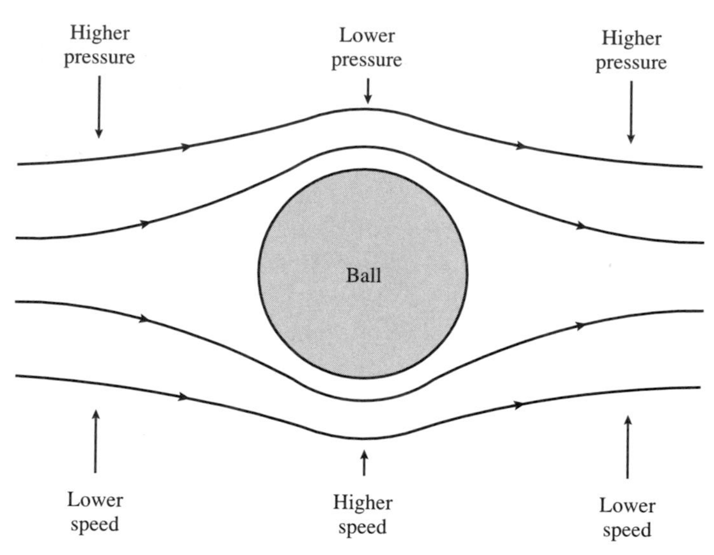

Figure 4.7. To maintain the flow where the channel between the streamlines narrows at the side of the ball, the air has to speed up. It slows again as the channel widens behind the ball. Pressure differences arise along the flow to drive the necessary acceleration and deceleration.

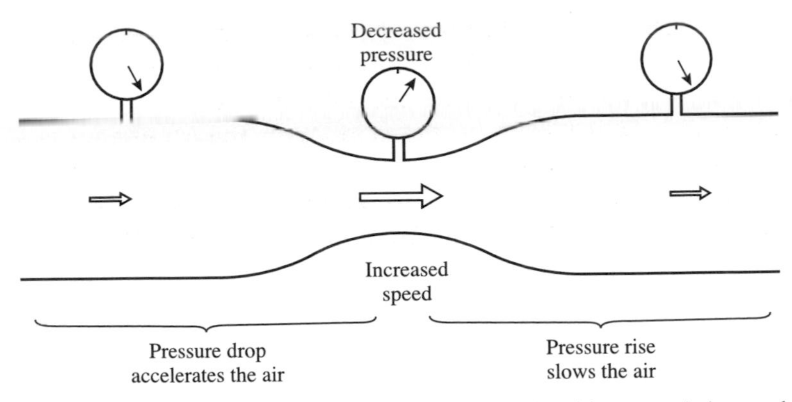

Figure 4.8. In this experiment air is passed down a tube with a constriction, and pressure gauges measure the pressure changes. The pressure falls as the flow speed increases, following Bernoulli's law.

has to move faster. The air speeds up as it approaches the side of the ball and then slows again as it departs at the rear.

For the air to be accelerated to the higher speed, a pressure difference arises, the pressure in front of the ball being higher than that at the side, the pressure drop accelerating the air. Similarly a pressure increase arises at the back of the ball to slow the air down again.

This effect can be seen more simply in an experiment where air is passed through a tube with a constriction as shown in figure 4.8. For the air to pass through the constriction it must speed up and this requires a pressure difference to accelerate the air. Consequently the pressure is higher before the constriction. Similarly the slowing of the air when it leaves the constriction is brought about by the higher pressure downstream. If pressure gauges are connected to the tube to measure the pressure differences they show a lower pressure at the constriction, where the flow speed is higher.

Separation of the flow

Why does the flow separate from the surface of the ball? As we have seen, the air is first accelerated and then decelerated but,

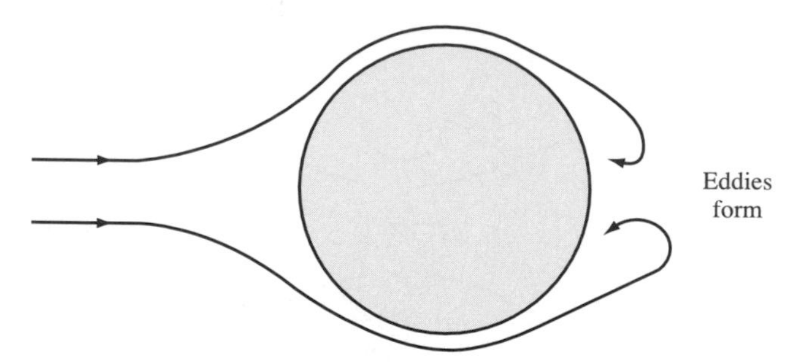

Figure 4.9. Viscosity slows the separated airflow, producing eddies behind the ball.

in addition to this, viscosity slows the air. As a result, the flow around the surface is halted towards the rear of the ball, and the flow separates from the surface.

This effect has been compared with that of a cyclist free-wheeling down a hill. His speed increases until he reaches the valley bottom. If he continues to free-wheel up the other side the kinetic energy gained going down the hill is gradually lost, and he finally comes to rest. If there were no friction he would reach the same height as the starting point, but with friction he stops short of this.

Similarly, the air in the boundary layer accelerates throughout the pressure drop and then decelerates throughout the pressure rise. Viscosity introduces an imbalance between these parts of the flow, and the air fails to complete its journey to the back of the ball. Figure 4.9 shows how the forward motion of the air is slowed and the flow turns to form an eddy.

The turbulent wake

The flow beyond the separation is irregular. Figure 4.10 illustrates the turbulent eddies which are formed, these eddies being confined to a wake behind the ball. The eddies in the flow have kinetic energy, and this energy has come from the loss of energy in the slowing of the ball.

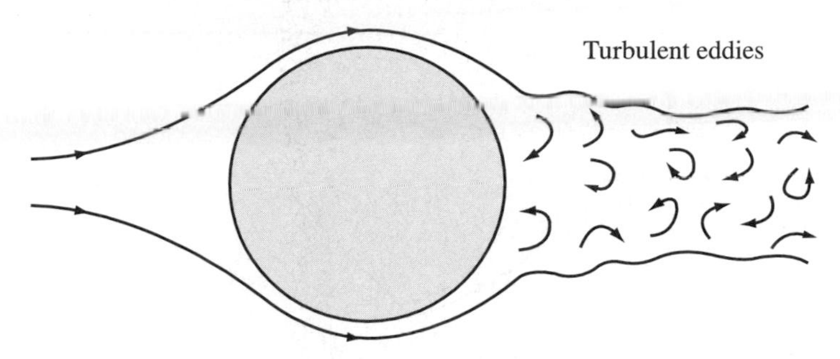

Figure 4.10. The separated flow is unstable and forms a turbulent wake.

With increasing ball speed the drag initially increases as the square of the speed, doubling the speed producing four times the drag. However, with further increase in speed there is a surprising change, and above a certain critical speed the drag force behaves quite differently.

The critical speed

There have been precise experimental measurements of the drag on smooth spheres. This allows us to calculate the drag force on a smooth sphere the size of a football, and the result is shown in figure 4.11. It is seen that there is an abrupt change around 50 miles per hour, a critical speed which is clearly in the speed range of practical interest with footballs. Above this critical speed the drag force actually falls with increasing speed, dropping to about a third of its previous value at a speed just over 60 miles per hour before increasing again.

However, although a football is smooth over most of its surface, the smoothness is broken by the stitching between the panels. Again surprisingly, the indentation of the surface caused by this stitching has a very large effect on the drag. There is little experimental evidence available on the drag on footballs, but measurements by the author indicate that the critical speed is much lower than for a smooth sphere, with

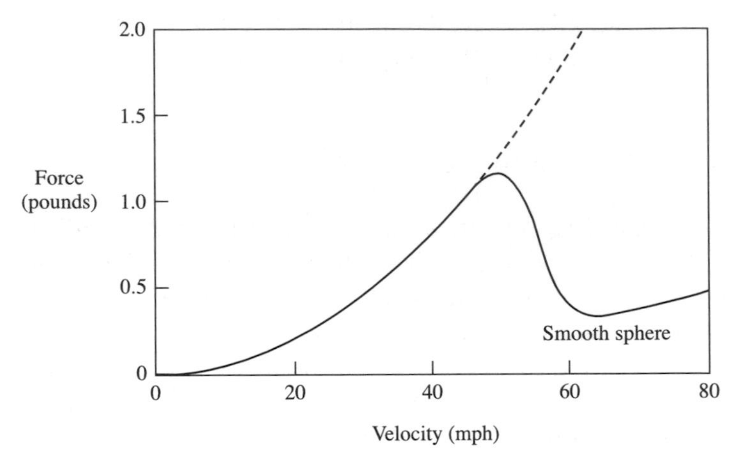

Figure 4.11. The graph shows how the drag force varies with speed for a smooth sphere the same size as a football. The dashed line gives a $(\text{speed})^2$ extrapolation.

a much less abrupt drop below the 'speed squared' line. Using these results, figure 4.12 shows how the drag on a football falls below that for a smooth sphere at low speeds and rises above it at high speeds. Also marked on the figure is the deceleration

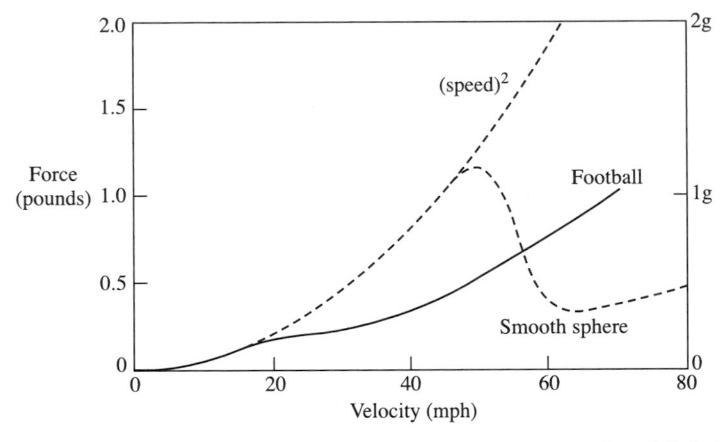

Figure 4.12. Drag force on a football. Above a critical speed the drag falls below the 'speed squared' dependence and below that for a smooth sphere. At high speeds the drag on the ball is greater than that for a smooth sphere. The deceleration which the drag force produces is shown on the right side in units of g.

which the drag force produces. With a deceleration of 1g the drag force is equal to the weight of the ball.

What happens at the critical speed?

Because the drag at low speeds is comparatively small, it is mainly for speeds above the critical speed that the flight of the ball is significantly affected by the drag. Our interest, therefore, is concentrated on these speeds.

The change in drag above the critical speed arises from a change in the pattern of the air flow. Above the critical speed the narrow boundary layer at the surface of the ball becomes unstable as illustrated in figure 4.13. This allows the faster moving air outside the boundary layer to mix with the slower air near the surface of the ball, and to carry it further toward the back of the ball before separation occurs. The result is a smaller wake and a reduced drag.

The onset of instability in the boundary layer around a sphere depends on the roughness of the surface. Rougher surfaces produce instability at a lower speed and consequently have a lower critical speed. A well-known example of this is

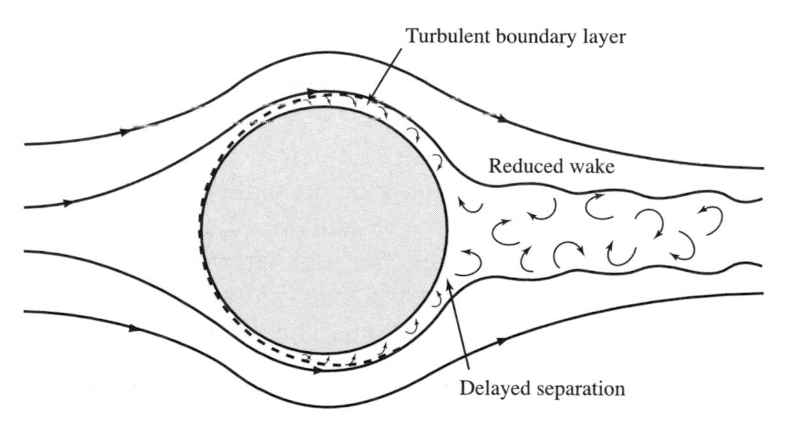

Figure 4.13. Above the critical speed the boundary layer becomes turbulent and this delays the separation, reducing the wake and the drag.

the dimpling of the surface of golf balls. Dimpling was introduced when it was found that initially-smooth golf balls could be driven further as their surface became rougher. The dimpling deliberately lowers the critical speed, reducing the drag in the speed range of interest, and allowing longer drives. With footballs the indentations along the stitching play a similar role, lowering the speed for the onset of instability in the boundary layer. At higher speeds the effect of roughness is to increase the drag above that for a smooth sphere.

Speed and range

There are two situations where players need to kick the ball at high speed. The first is when a striker or a penalty-taker has to minimise the time the goalkeeper has to react and launch himself toward the ball. In a penalty-kick the ball reaches the goal in a fraction of a second and in this brief time air drag only reduces the speed of the ball by about 10%.

The objective of a fast penalty kick is to put the ball over the goal-line before the goalkeeper can reach it. The time it takes for a ball to travel from the foot to cross the goal-line is given by the distance travelled divided by the speed of the ball. Provided accuracy is maintained, the faster the kick the better. With this objective, penalty-takers achieve ball speeds up to 80 miles per hour.

The distance of the penalty spot from the goal-line is 12 yards. In a well-struck penalty kick the ball travels further to the goal, being aimed close to the goal post, but never needing to travel more than 13 yards to the goal. An 80 miles per hour penalty kick travels at 39 yards per second and so its time of flight is about a third of a second. This is comparable with the reaction time of a goalkeeper, and so the only chance a goalkeeper has with a well-struck penalty kick is to anticipate which side the ball will go and use the one third of a second diving through the air.

The second type of kick which needs a high speed is the long kick. In particular, the goalkeeper is often aiming to achieve maximum range, whether kicking from his hand or from the six-yard box. In the absence of air drag the distance reached would increase as the square of the initial speed, twice the speed giving four times the range. Because of air drag this doesn't happen. At higher speed the drag is more effective in reducing the speed during the flight of the ball, and we shall find that this greatly reduces the range.

A goal-kick can be kicked at a similar speed to a penalty shot but, because of the longer time of flight, the air drag significantly affects its path. For a well-struck kick with a speed of 70 miles per hour, the force due to the drag is about the same as the force due to gravity. The range of a kick in still air is determined by the initial speed of the ball and the initial angle to the horizontal. For a slow kick the effect of drag is negligible. In that case there is practically no horizontal force on the ball, and the horizontal part of the velocity is constant in time. For high speed kicks the air drag rapidly reduces the speed of the ball, as illustrated in figure 4.14 which shows the fall in the horizontal velocity for a 70 mile per hour kick.

The range depends on the average horizontal velocity of the ball, and on the time of flight. Both of these factors are reduced by air drag, the fall in horizontal velocity having the larger affect. Figure 4.15 shows how the range depends on the initial speed for a kick at 45°. In order to bring out the effect of air drag, the range calculated without air drag is shown for comparison. It is seen that, for high speed kicks, air drag can reduce the range by half.

The effect of air drag on the path of the ball is illustrated in figure 4.16, which shows the flight of a 70 miles per hour kick at 45°. The drag reduces both the vertical and the horizontal velocities but the greater effect on the horizontal velocity means that the ball comes to the ground at a steeper angle than that of the symmetric path which the ball would take in the absence of drag. When air drag is allowed for, it

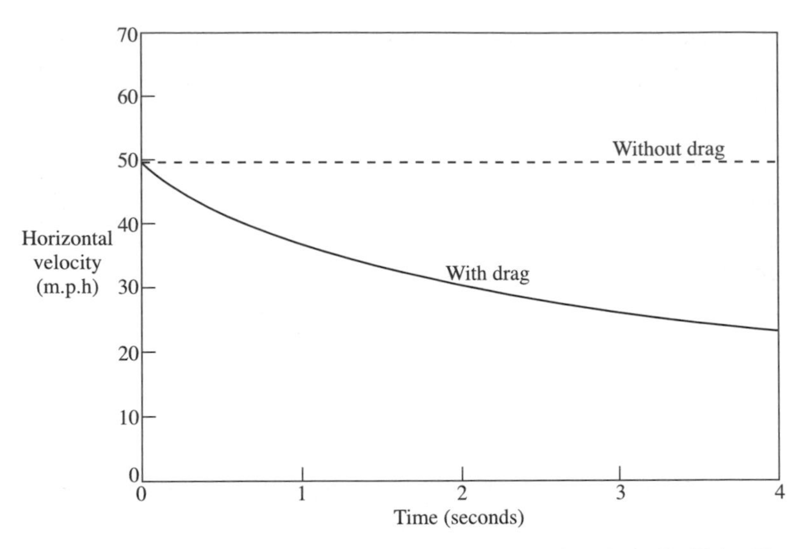

Figure 4.14. The drag on the ball reduces its velocity during the ball's flight. The graph shows the fall in the horizontal part of the velocity with time for a 70 miles per hour kick at 45°. The horizontal velocity starts at 50 miles per hour and is roughly halved by the time the ball reaches the ground.

turns out that 45° no longer gives the maximum range for a given speed. Because the main effect of the drag is to reduce the horizontal velocity, the maximum range is obtained by making some compensation for this by increasing the initial horizontal velocity at the expense of the vertical velocity. This means that the optimum angle is less than 45°. Although at high speeds the optimum angle can be substantially lower than 45°, it turns out that the gain in range with the lower angle is slight, typically a few yards.

Nevertheless goalkeepers do find that they obtain the longest range goal-kicks with an angle lower than 45°, but this might be unrelated to air drag. The reason possibly follows from the fact that the achievable speed depends on the angle at which the ball is kicked. The mechanics of the kick are such that it is easier to obtain a high speed with a low angle than a high angle. Just imagine trying to kick a ball vertically from the ground.

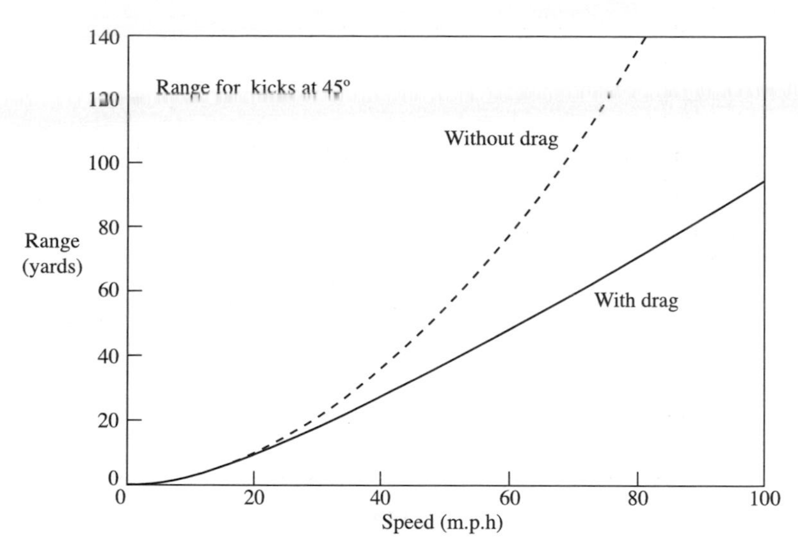

Figure 4.15. For balls kicked at a given angle the range depends only on the speed with which the ball is kicked. The graph shows the dependence of range on the initial speed for kicks at 45°. The range calculated without air drag is given for comparison.

Generally long range goal-kicks are kicked at an angle closer to 30° and a typical goal-kick lands just beyond the centre circle. The speed needed for a given range has been calculated and it can be seen from figure 4.17 that such a goal-kick requires an initial speed of 70 miles per hour. The calculation also gives the time of flight of the ball, and the dependence of this time on the range is shown in figure 4.18.

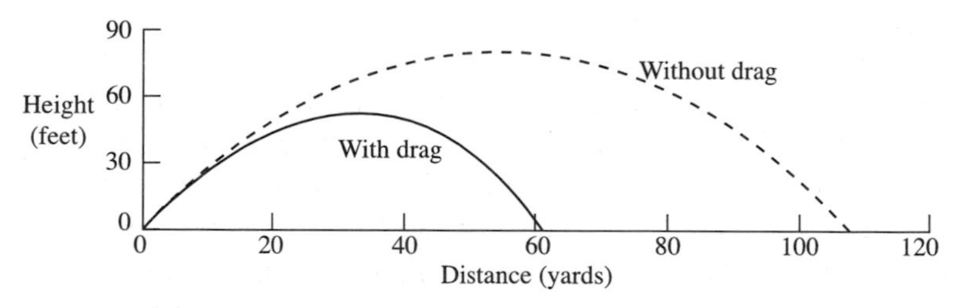

Figure 4.16. Path of ball kicked at 70 miles per hour and 45°. Comparison with the path calculated without air drag shows the large effect of the drag.

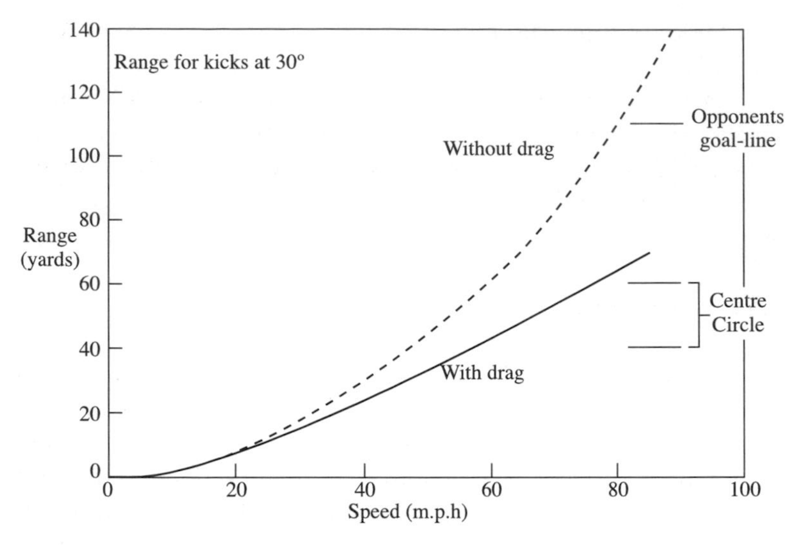

Figure 4.17. Range calculated for kicks at 30° to the horizontal.

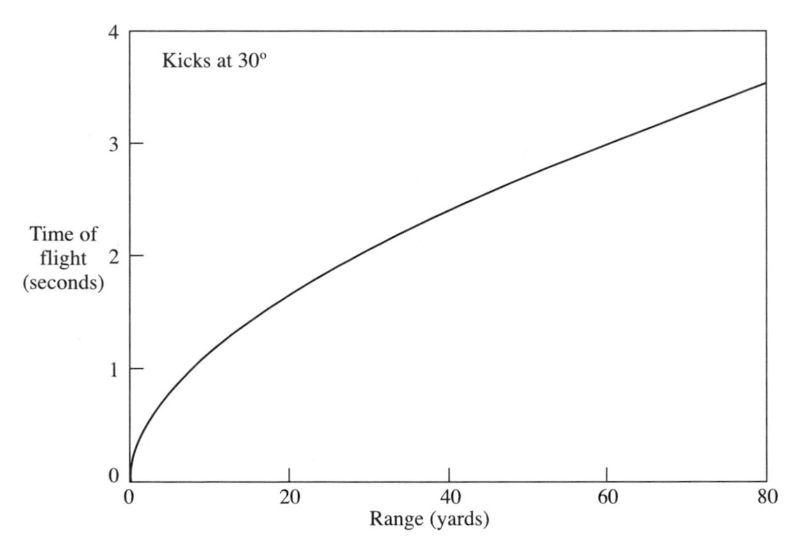

Figure 4.18. The graph shows how the time of flight increases with range for balls kicked at 30°.

Most long range goal-kicks have a time of flight of about 3 seconds.

Balls kicked after being dropped from the goalkeeper's hands are easier to kick at a higher angle than are goal-kicks, and generally goalkeepers do make such kicks at an angle closer to 45°.

Effect of a wind

When there is a wind, the speed of the air over the ball is changed and there is an additional force on the ball. This force depends on the speed of the ball and is approximately proportional to the speed of the wind. It is clear that a tail wind will increase the range of a kick and a head wind will decrease the range. For a goal-kick, a rough approximation is that the range is increased or decreased by a yard for each mile per hour of the wind. For example a goal kick which without a wind would reach the back of the centre circle, would be carried by a 30 mile per hour tail wind into the penalty area. It is kicks of this sort which occasionally embarrass the goalkeeper who comes out to meet the ball, misjudges it, and finds that the bounce has taken it over his head into the goal.

A strong head wind can seriously limit the range. Figure 4.19 shows the path of the ball in two such cases. The first is for a 70 miles per hour kick into a 30 miles per hour head wind. It is seen that the forward velocity is reduced to zero at the end of the flight, the ball falling vertically to the ground. The second is that for an extreme case with a 40 miles per hour gale. The horizontal velocity is actually reversed during the flight, and the ball ends up moving backwards.

When there is a side wind the ball suffers a deflection. As we would expect, this deflection increases with the wind speed and with the time of flight. A 10 miles per hour side wind displaces the flight of a penalty kick by a few inches. This is unlikely to trouble a goalkeeper but a 1 foot deflection in a

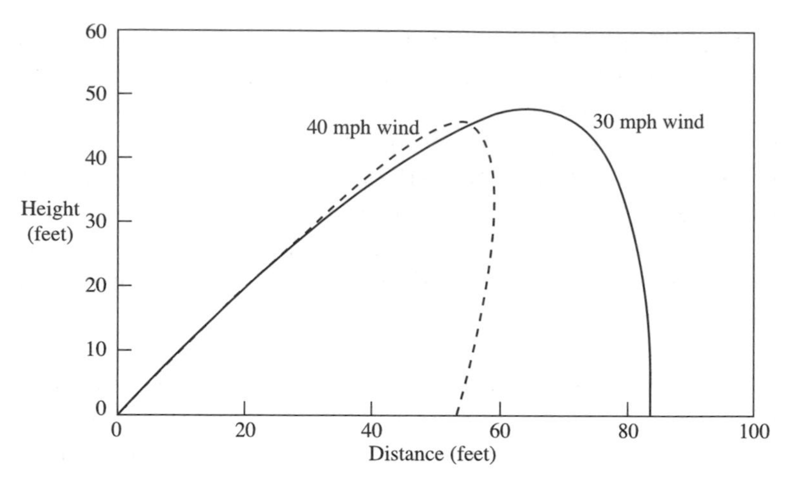

Figure 4.19. The effect of a strong head wind on the paths of 70 miles per hour kicks at 45°.

30 miles per hour wind might, especially as the wind causes the flight to be curved.

A 10 miles per hour side wind would deflect a 20 yard kick by about a yard, and a goal kick by about 5 yards. It is clear from this how games can be spoilt by strong winds, especially if gusty. Players learn to anticipate the normal flight of the ball, and there is some loss of control when the ball moves in an unexpected way.

The banana kick

The simple theory of the flight of the ball predicts that, in the absence of wind, the ball will move in a vertical plane in the direction it is kicked. It is surprising, therefore, to see shots curling on their way to the goal. The same trick allows corner kicks to cause confusion in the defence by either an inward- or outward-turning flight of the ball.

Viewed from above a normal kick follows a straight line. This is consistent with Newton's law of motion which tells us that the appearance of a sideways movement would require a sideways force. We see, therefore, that to understand the curled flight of a ball we must be able to identify and describe this sideways force.

The first clue comes from the kicking of the ball. To produce a curled flight the ball is not struck along the line of its centre. The kick is made across the ball and this imparts a spin. It is this spin which creates the sideways force, and the direction of the spin determines the direction of the curve in flight.

Attempts to explain the curved flight of a spinning ball have a long history. Newton himself realised that the flight of a tennis ball was affected by spin and in 1672 suggested that the effect involved the interaction with the surrounding air. In 1742 the English mathematician and engineer Robins explained his observations of the transverse deflection of musket balls in terms of their spin. The German physicist Magnus carried out further investigations in the nineteenth century, finding that a rotating cylinder moved sideways when mounted perpendicular to the airflow. Given the history, it would seem appropriate to describe the phenomenon as the Magnus–Robins effect but it is usually called the Magnus effect.

Until the twentieth century the explanation could only be partial because the concepts of boundary layers and flow separation were unknown. Let us look at the simple description of the effect suggested in earlier days. It was correctly thought that the spinning ball to some extent carried the air in the direction of the spin. This means that the flow velocity on the side of the ball moving with the airflow is increased and from Bernoulli's principle the pressure on this side would be reduced. On the side moving into the airflow the air speed is reduced and the pressure correspondingly increased. The resulting pressure difference would lead to a force in the observed direction. However, this description is no longer acceptable.

With the understanding that there is a thin boundary layer around the surface comes the realisation that the viscous

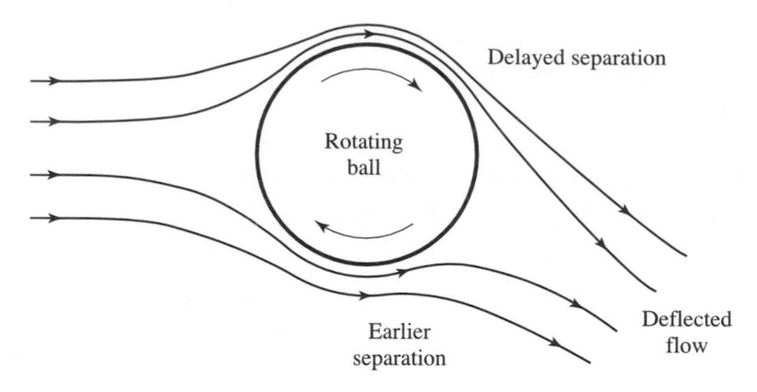

Figure 4.20. Rotation of the ball leads to an asymmetric separation.

drag on the air arising from the rotation of the ball is limited to this narrow layer, and because of the viscous force in this layer Bernoulli's principle does not hold.

There are two steps to an understanding of what actually happens with a spinning ball. The first is to see the pattern of flow over the ball and the second is to understand how this implies a sideways force.

We saw earlier how, with a non-spinning ball, the air flows over the surface of the ball until it is slowed to the point where separation occurs. With spin an asymmetry is introduced as illustrated in figure 4.20. On the side of the ball moving with the flow the viscous force from the moving surface carries the air farther around the ball before separation occurs. On the side of the ball moving against the flow the air is slowed more quickly and separation occurs earlier. The result of all this is that the air leaving the ball is deflected sideways.

We can see from the flow pattern that the distribution of air pressure over the ball, including that of the turbulent wake, will now be rather complicated. There is, therefore, no simple calculation which gives the sideways force on the ball. However, we can determine the direction of the force. The simplest way is to see that the ball deflects the air to one side and this means that the air must have pushed on the ball in the opposite direction as illustrated in figure 4.21. In more

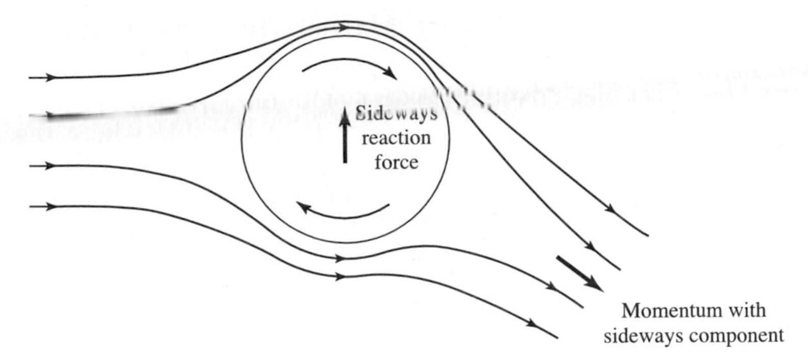

Figure 4.21. Airflow is deflected by spin, with a sideways reaction force on the ball.

technical terms the sideways component of the airflow carries momentum in that direction and, since the total momentum is conserved, the ball must move in the opposite direction taking an equal momentum. This is the Magnus effect.

Having determined the direction of the force we can now work out the effect of spin on the flight. In figures 4.20 and 4.21 the airflow comes to the ball from the left, meaning that we have taken the motion of the ball to be to the left. The direction of the Magnus force is then such as to give the curved flight shown in figure 4.22. If the spin imparted at the kick were in the other direction the ball would curve the other way.

With a very smooth ball, like a beach-ball, a more irregular sideways motion can occur. The ball can move in the opposite direction to the Magnus effect and can even undergo sideways shifts in both directions during its flight.

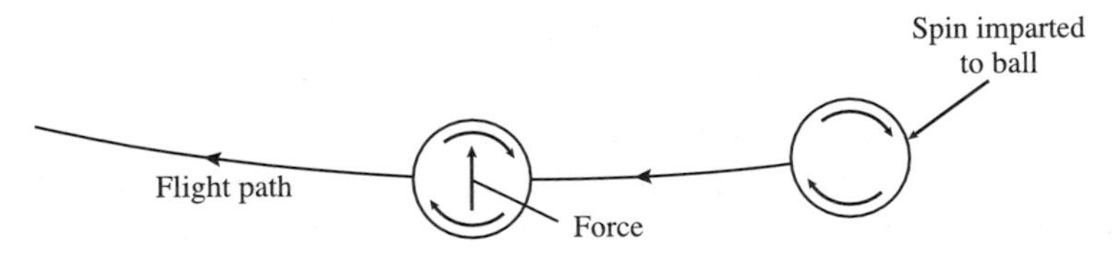

Figure 4.22. Showing the direction the ball curls in response to the direction of the spin.

We can see how an inverse Magnus effect can occur by recalling that there is a critical speed above which the boundary layer becomes unstable. With a spinning ball the air speed relative to the ball's surface is higher on the side where the surface is moving against the air. We would, therefore, expect that over a range of ball speeds the critical flow speed can be exceeded on this side of the ball and not exceeded on the other. Since the effect of the resulting turbulence is to delay separation, we see that the asymmetry of the flow pattern can now be the opposite of that occurring with the Magnus effect and the resulting force will also be reversed.

The more predictable and steady behaviour of a good football must be due to a more regular flow pattern at the surface of the ball initiated by the valleys in the surface where the ball is stitched.

Chapter 5

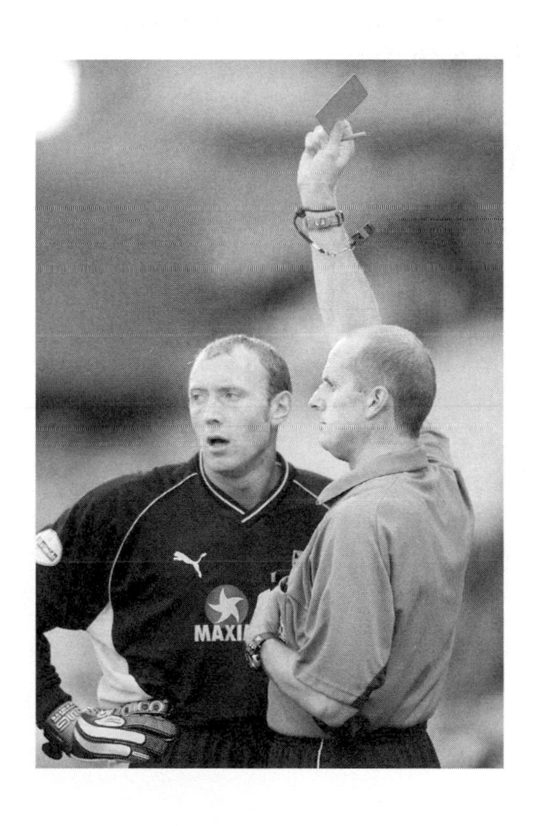

5

The laws

Football was first played with codified rules in the middle of the nineteenth century. Although the game bore some relation to the modern game there were fundamental differences. For example in the early games the ball could be handled as in rugby, and 'hacking' was allowed. One dominant concept was that the ball should be 'dribbled' forward and that players should keep behind the ball. Later, forward passing was allowed but the idea that the ball should be worked forward persists in the present off-side rule.

Initially there was a variety of rules, each school or club being free to decide for itself. The growth of competition demanded a uniform set of rules and by 1870 'soccer' was completely separated from rugby and was recognisable as the modern game.

The process by which the present laws emerged was of course empirical. The laws were refined to improve the game for both players and spectators. However, this does not mean that no principles are involved and we can ask why the laws have their present form. Of course the issues are complex and the laws are interdependent, so we cannot expect simple answers. Nevertheless it is of interest to try to uncover some of the underlying principles.

To take an example, we can ask why the goals are the size they are – 8 feet high, 8 yards wide. The basic determining factor is the number of goals desirable in a match. If the goal were twice as wide the scoring rate would be phenomenal, and if it were half as wide there would be a preponderance of 0–0 draws. So the question becomes what is the optimum scoring rate, or goals per match, and we shall return to this later.

Further questions are why the pitch is the size it is, and why eleven players? In the early days the pitch would be whatever piece of land was available but it would soon be clear that it would best be large enough that the goal could not be bombarded by kicks from the whole of the pitch. In more recent times commercial factors demand that the pitch be a suitable size for the spectators. However, it is probably a coincidence that the chosen size of the pitch allows even the largest number of spectators to be accommodated with a reasonable view of the game. The question of how many players leads to an even more basic question as to whether there is a relationship between the various fundamental factors involved. If there is such a relation this might provide the starting point for a 'theory of football'. Let us now examine this question.

With respect to the play there must be a general relation between the number of players and the best size of the pitch, six-a-side matches obviously needing a smaller pitch. It seems likely that the essential factor is that there be pressure on the players to quickly control the ball and decide what to do with it. This means that opposing players must typically be able to run to the player with the ball in a time comparable with the time taken to receive, control and move the ball. If the distance between players is larger the game loses its tension. If this distance is much less the game has the appearance of a pin-ball machine. We cannot expect to be able to do a precise calculation, but we can carry out what is often called a back-of-envelope calculation to see the rough relationship between the quantities involved and to check that the numbers make sense.

If there are N outfield players in each team and the area of the pitch is A, the number of these players per unit area is

$n = N/A$. A simple calculation gives the average distance to the nearest opponent as approximately $d = \frac{1}{2}/\sqrt{n}$. If the speed with which players move to challenge is s, the time to challenge is d/s. Thus, if the time to receive, control and decide is t and we equate this to the time to challenge, we obtain the optimal relationship between the four basic factors t, A, s and N as

$$t \simeq \frac{1}{2s}\sqrt{\frac{A}{N}}$$

where the symbol $\simeq$ indicates the lack of precision in the equality. Taking the area of the pitch to be 110 yards $\times$ 70 yards $=$ 7700 square yards and the speed of the players as 5 yards/second we obtain $t \simeq 9/\sqrt{N}$ and figure 5.1 gives the corresponding plot of t against N. We see that for $N = 10$, as specified by the rules, the characteristic time has the quite

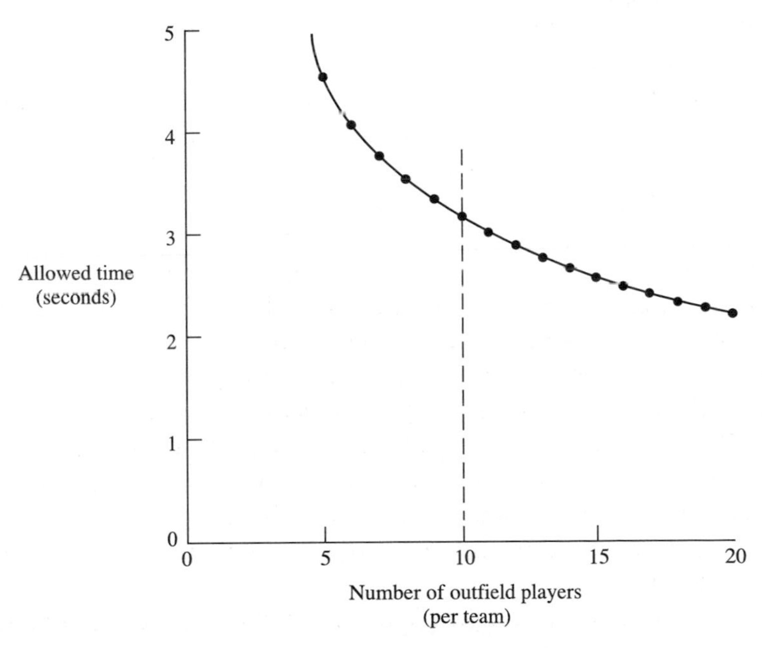

Figure 5.1. The allowed time depends on the number of players.

reasonable value of 3 seconds. This might typically allow a second for the pass a player receives, a second to control and a second to either release the ball or start running with it.

How many goals?

Perhaps the most frequently raised issue concerning the laws is whether the number of goals scored in a match should be increased. The number could easily be adjusted, for example, by changing the height and width of the goal. This leads us to ask what factors are involved in deciding the optimum number of goals per match.

That there is an optimum is clear. Obviously zero goals is no good and, on the other hand, no-one wants to see basketball scores. Since both of these limits are completely unsatisfactory there has to be an optimum in between.

Basically, very low scoring is not acceptable because we miss the excitement of goals being scored. This is particularly true of 0–0 draws which are generally regarded as disappointing.

The case against high scoring is less clear. In basketball and rugby high scores are found quite acceptable. One argument is that the larger the number of goals, the less significant and exciting is each goal. Another is that the results of matches become more predictable. With opposing teams of equal ability both teams have an equal chance of winning no matter what the average scoring rate, but for teams of unequal ability the average scoring rate matters. As we shall see, the weaker team has a better chance of providing an 'upset' if the scoring is lower. This must be regarded as an argument against a high scoring rate because the enjoyment is reduced if the result is predictable and the better team almost always wins. We shall shortly examine the reason why the weaker team benefits from a lower scoring rate, but in order to do so we need to introduce the concept of probability.

Probability is measured on a scale of 0 to 1, zero applying to impossibility and 1 to certainty. Thus a probability of 1 in 4 is 0.25 and 1 in 2 is 0.5 and so on. It is sometimes convenient to express the probability as a percentage, thus 0.25 and 0.5 become 25% and 50% for example. In considering the probabilities of the various outcomes we know that, since there must be some outcome, the sum of the probabilities of all possible outcomes will be 1.

We now return to the effect of the scoring rate on the chance of the weaker team winning. This can be illustrated by considering matches in which the better team has twice the potential scoring rate of its opponent. The probability of the weaker team winning depends on whether the total number of goals scored is odd or even, a draw being impossible with an odd number of goals. First we look at matches with an odd number of goals.

If only one goal is scored, the probability that it is scored by the stronger team is 2/3 and the probability that is scored by the weaker team is 1/3. The weaker team has, therefore a 33% chance of being the winner.

With three goals the situation is more complicated. We must take account of the possible orders of goal scoring and calculate the probability of each. If the weaker team wins 3–0 there is only one possible sequence of three goals, which we can write *www* where *w* denotes a goal by the weaker team. The probability of this sequence is $\frac{1}{3} \times \frac{1}{3} \times \frac{1}{3} = \frac{1}{27}$. For a 2–1 win for the weaker team there are three possible sequences. Denoting a goal by the stronger team by *s* these are *wws*, *wsw* and *sww*. The probability of each of these sequences with two goals to the weaker team and one to the stronger is $\frac{1}{3} \times \frac{1}{3} \times \frac{2}{3} = \frac{2}{27}$, so allowing for the three possible sequences the probability of a 2–1 win for the weaker team is $3 \times \frac{2}{27} = \frac{6}{27}$. Since 3–0 and 2–1 are the only scores for a win, the total probability of a win for the weaker team is $\frac{1}{27} + \frac{6}{27} = \frac{7}{27}$ or 26%. We see that with three goals as compared with one goal the probability of the weaker team winning is reduced from 33% to 26%.

As the number of goals in the match increases the probability of the weaker team winning continues to fall.

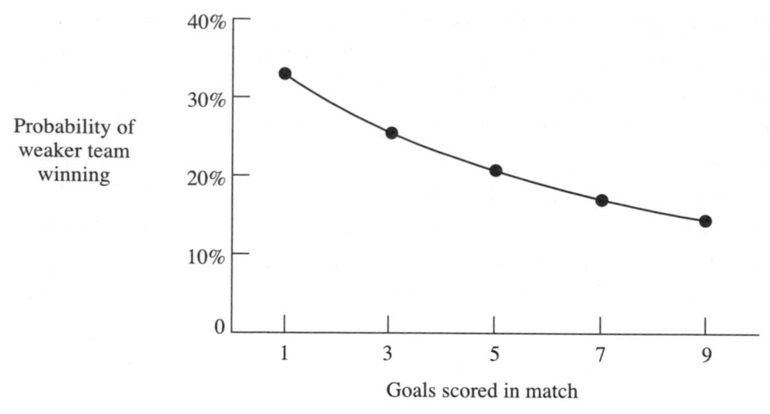

Figure 5.2. The probability of the weaker team winning depends on the total number of goals scored in the match. The graph shows the dependence when the number of goals is odd.

Figure 5.2 gives a graph showing the probability of a win for each number of goals. At nine goals it has fallen below 15%.

Similar calculations with an even number of goals scored in the match give the results shown in figure 5.3, which also

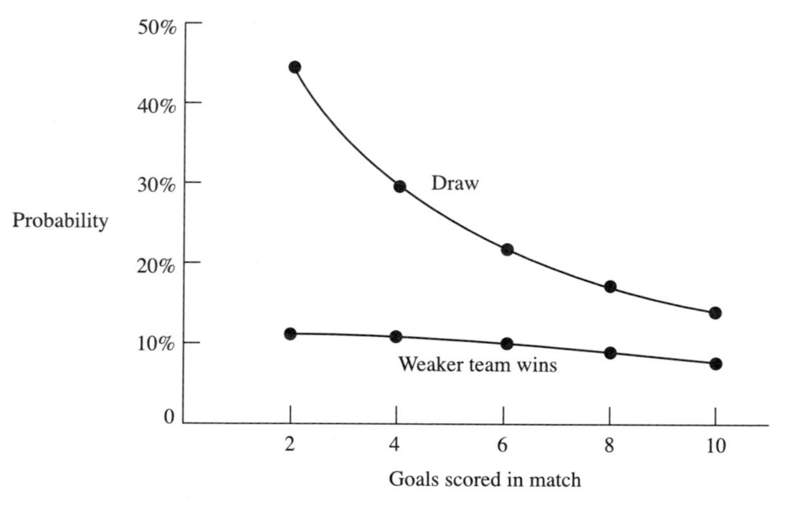

Figure 5.3. Probability of a draw and a win for the weaker team when the total number of goals is even.

includes the probability of a draw. It is seen that with an even number of goals the reduction in the weaker team's chance of winning as the total number of goals is increased is only slight. However, the chance of coming away with a draw falls very rapidly.

The choice of a two to one scoring ratio in the above example is, of course, arbitrary. It does, however, illustrate an important advantage of the rules not allowing too high a scoring rate. The excitement from the uncertainty as to the outcome with the improved chance of the weaker team getting a surprise result outweighs the occasional 'injustice' to the stronger team.

Imprecision of the laws

Some imprecision in the laws of a game may be valuable if it allows the referee or umpire to use his common sense. In the case of football the imprecision is sometimes unhelpful or unnecessary.

The off-side law is such a case. The law states that a player shall not be declared off-side by the referee merely because of being in an off-side position. He shall only be declared off-side if, at the moment the ball touches or is played by one of his team, he is in the opinion of the referee (a) interfering with play or with an opponent, or (b) seeking to gain an advantage by being in that position.

The use of the phrase 'interfering with play' is rather mysterious. Presumably it is influencing the play which is precluded. Regarding (b), even if the player is not gaining an advantage from being where he is, it seems a curious idea that he is not *seeking* an advantage, and if he is seeking an advantage surely he is influencing the play.

The problem is actually deeper, for if we allow that an attacking player is not 'interfering' and not seeking an advantage, his intentions may not be clear to the defenders, whose positioning and attention are then affected. This means that

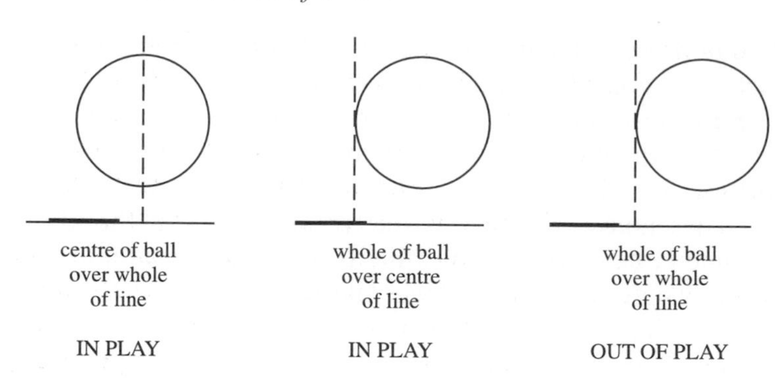

Figure 5.4. The ball is out of play when the whole of the ball has crossed the whole of the line.

a player whose intentions are benign can nevertheless influence the play. It is not clear how the referee is supposed to assess all of this in the brief time available.

A minor irritation in football is the imprecision with which the law relating to the ball being out-of-play is applied by linesmen. Whether this is due to vagueness as to the rule, or carelessness in its application, is not clear. The law states that the ball is out of play 'when it has wholly crossed the goal-line or touch-line, whether on the ground or in the air'. Linesmen often seem to be interpreting 'wholly' as meaning 'the whole of the ball over the centre of the line' or 'the centre of the ball over the whole of the line'.

The law should actually read 'The ball is out of play when the whole of the ball has crossed a vertical plane containing the outside edge of the line'. More simply, but less precisely, the ball is out of play when the whole of the ball has crossed the whole of the line. The various cases are illustrated in figure 5.4.

Free-kicks

Free-kicks are partly a deterrent against unacceptable play, and partly a compensation to the aggrieved team for the loss of opportunity arising from the infringement of the rules. The present law regarding free-kicks seems to be generally accepted as satisfactory. One reason for this is that they contain an implicit variation of significance according to the position on the field. An infringement by a team in its opponent's half of the pitch does not usually affect their opponent's chances a great deal, and the value of the resulting free-kick to the opponents is appropriately small. On the other hand an infringement 20 yards out from the goal by the defending team can mean a substantial loss of opportunity to the attacking team, and the resulting free-kick provides the proper compensation of a useful shot on goal.

Penalties

The award of a penalty-kick is almost, but not quite, the same as the award of a goal. The probability of a goal being scored from a penalty kick is typically 70 to 80% depending, of course, on the penalty-taker. Penalty-kicks provide only a rough form of justice. Sometimes a marginal handling offence leads to a penalty-goal, whereas a penalty-kick awarded for illegally preventing an almost certain goal can fail. The uncertainty of penalties actually contributes to the excitement of the game.

The strategy of the penalty-taker is to aim the shot wide of the goalkeeper but sufficiently clear of the goal-post to allow for a range of error. Until 1997 the goalkeeper was constrained to keep his feet still on the goal-line until the ball was kicked. The rule was then changed to allow the keeper to move, but only along his line. Clearly the goalkeeper's best strategy is to give himself a chance by guessing which side of him the ball will be placed, and to start his initial movement before the ball is struck. On the other hand he must not start so early as to betray his choice to the penalty-taker.

The high scoring rate from penalties is implicit in the rules. The choice of 12 yards for the distance of the penalty spot from the goal-line clearly implies a judgement as to what is fair. The average scoring rate from penalties could be adjusted by altering the distance of the penalty spot.

If the distance were zero, the penalty-kick being taken from the goal-line, the goalkeeper could obviously block the shot by standing behind the ball. Indeed the introduction of penalty-kicks in 1891 was very much influenced by the blocking of a free-kick on the goal-line in an F.A. Cup quarter-final. The free-kick had been awarded to Stoke when a Notts County defender punched the ball off the line to prevent an otherwise certain goal. The Notts County goalkeeper successfully blocked the free-kick, Stoke lost 1–0, and Notts County went through to the semi-final.

As the penalty spot is moved away from the goal-line it initially becomes easier to score, the scoring probability approaching certainty at a few yards. For larger distances the probability falls and at very large distances becomes zero. Figure 5.5, which is based on a session of experimental penalty-kicks taken by skilled players, gives an indication of what the scoring rate would be for different distances of the penalty spot.

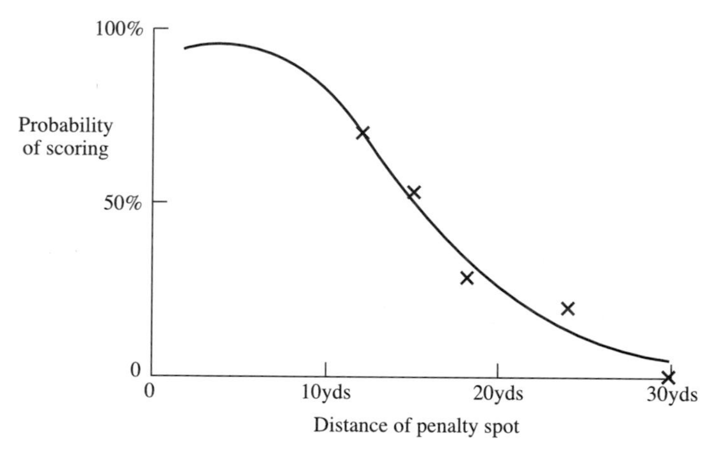

Figure 5.5. The probability of scoring from a penalty kick depends on the distance of the kick. The crosses mark the experimental results.

For a penalty spot distance of around 3 yards the scoring probability approaches 100% because the ball can be safely kicked at high speed beyond the goalkeeper's reach, but well away from the goal-post. For example, a 60 mile per hour low shot from 3 yards out, aimed 3 feet from the goal-post, would pass the goal-line 9 feet away from the goalkeeper in $\frac{1}{7}$ of a second, giving the goalkeeper virtually no chance. As we move the penalty spot further away the scoring probability begins to fall, reducing in the experimental case to a 70% rate for 12 yards and falling continuously as the distance is increased farther.

A top-class goalkeeper can cover the whole of the goal given a little more than a second. A good penalty taker can kick the ball at 80 miles per hour. This gives us an estimate of the maximum distance from which a penalty kick could be successful. Allowing for air drag, a perfectly taken penalty kick at 80 miles per hour driven into the top corner of the goal could defeat the goalkeeper from about 35 yards.

We see from the above analysis that the choice of 12 yards for the penalty spot implies a choice of scoring probability. However, the matter is rarely discussed and presumably this means that, taking all factors into account, the distance chosen in 1891 is about right.

Competitions

In addition to the question of the rules of the game, we can ask about the rules of competitions. Should we, for example, have penalty shoot-outs and 'golden goals'? Some care is needed in deciding the rules of competitions, as can be illustrated by the wonderful fiasco in a match between Barbados and Grenada. It was the final group match of the Shell Caribbean Cup and this is what happened.

A rule of the competition was that, in a match decided by a sudden-death 'golden goal' in extra time, victory would be deemed equivalent to a 2–0 win. Barbados needed to win by at least two goals to reach the finals. Otherwise Grenada qualified. The Barbados team was on its way midway through the second half, leading 2–0. However, Grenada pulled one back, making the score 2–1. If the score remained unchanged Barbados was out. With three minutes to go the Barbados team realised that they would be more likely to win in extra time than score the required goal in the remaining minutes. They therefore turned their attack on their own goal and scored, bringing the scores level at 2–2, with the consequent possibility of victory in extra time.

Grenada saw the point, and tried to lose the match, attempting to achieve qualification by scoring an own goal to make the score 3–2. However, Barbados sprang to the defence of the Grenada goal and kept the score at 2–2. After four minutes of extra time Barbados scored the golden goal and qualified for the finals.

Chapter 6

6

Game theory

Football is the best of games. Its superiority derives from two sources, variety and continuity. At each point in the game the players are faced with a wide range of options – take the ball past the opponent on this side or that, to pass – short or long, low or high, to shoot – or to lay the ball off – and to whom. Compared with other games the flow of the game is continuous, the ball being in play for most of the time. Even the delays for free kicks and corner kicks add to the excitement and penalty kicks are often times of high drama.

The richness of the game makes it difficult to give a theoretical description. The unexpected, imaginative touches which are crucial to the game defy a theoretical approach. However, it is often the case in science that by giving up any attempt to include the detail, and allowing as much simplification as possible, a description of the broader features of a subject can be achieved. This is also the case with football.

Random motion?

At any time during a match the play (one hopes) appears purposeful. But if we take a bird's eye view of the motion of the ball it has the appearance of random motion. Figure 6.1 shows the movement of the ball during the six minutes

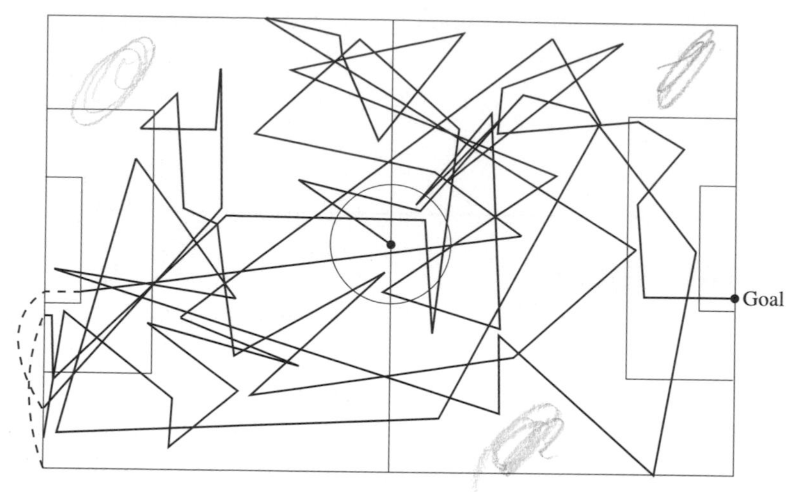

Figure 6.1. Movement of the ball over the pitch in a European Championship match between England and Holland.

between Sheringham's first goal and Shearer's second in the 1996 European Championship match between England and Holland. The behaviour of the ball is reminiscent of a phenomenon called Brownian motion. It was noticed by the Scottish botanist Robert Brown that, when viewed under a microscope, pollen grains suspended in water are seen to undergo erratic motion. The theory of this behaviour was provided by Einstein in terms of the impact of the water molecules on the suspended pollen grains.

In the case of football the strength and deployment of the team is the factor which moderates the random motion. For example, with unequal teams the ball spends more time in the weaker team's half and with two defensive teams the ball becomes trapped in midfield. These two cases are illustrated in figure 6.2 in which the randomness is averaged out to give graphs of the average time spent in each part of the pitch.

A proper theoretical treatment would call for quite sophisticated techniques and no such theory has been developed. However, some introductory thoughts are discussed in chapter 10.

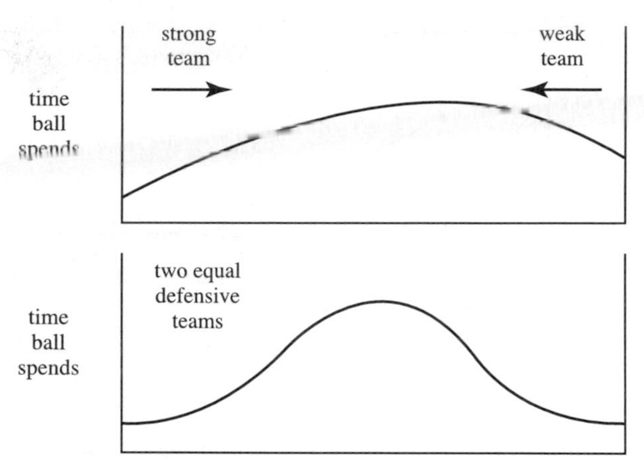

Figure 6.2. Distribution of time spent over the length of the pitch.

Scoring

We now look at the scoring during a game. The simplification we shall allow is that each team has an average scoring rate against the type of opponent they face. For a particular team the average scoring rate can be derived by taking the total number of goals scored against similar standard opposition over several games and dividing by the total playing time. For simplicity we first consider a match with one team having an average scoring rate of 1 goal per hour. With the chosen scoring rate the probability of the team scoring a goal in the first minute is 1 in 60. After 5 minutes the probability of having scored a goal is approximately 1 in 12 – 'approximately' because we cannot just add probabilities. We have to be more careful and also take account of the possibility of 2 or more goals being scored. It is possible to calculate the probability for each number of goals, and the results are shown in figure 6.3. Since at all times it is certain that the team has scored *some* number of goals (including zero) the probabilities of each number of goals must add up to 1.

Examining the figure we see that, as we would expect, at the outset the probability of zero goals is 1, it being certain

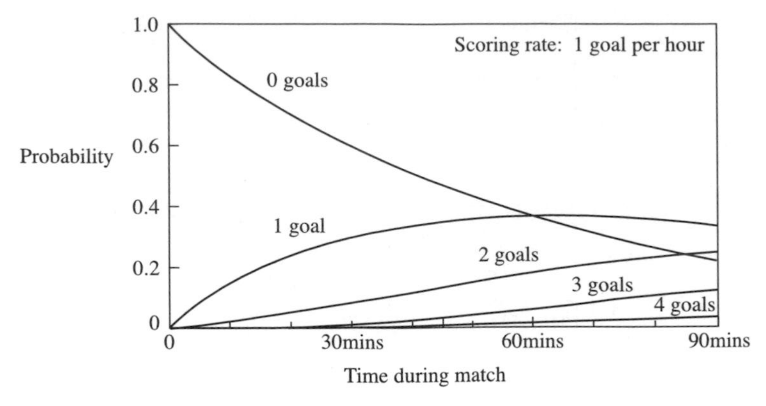

Figure 6.3. Probability of number of goals scored during a match for a team with an average scoring rate of one goal per hour.

that no goals have been scored. As time goes on the probability that the team has scored no goals falls, reaching 0.22 after 90 minutes. So, with the chosen rate of 1 goal per hour, there is just over a 1 in 5 chance that the team would not score. In the Premiership the average probability of not scoring in a match is about 1 in 4. Correspondingly, the probability that the team *has* scored increases with time. At half-time the probability that they have scored just 1 goal is 0.35. After an hour the probability that the team has scored just 1 goal begins to decrease reaching 0.33 at full time. The reason for the fall, of course, is the increasing likelihood that the team has scored more goals. At the end of the game it is more likely that they have scored more than 1 goal, than only 1 goal.

Let us now imagine that the team is playing a somewhat weaker opponent with an average scoring rate of a goal every 90 minutes. Again we can calculate the probability of this team having scored any number of goals at each time. The result is shown in figure 6.4. We see that the most likely score for this team is zero throughout the match, with an equal likelihood of 1 goal at full time. This doesn't mean, of course, that the stronger team will necessarily win, and we can use the

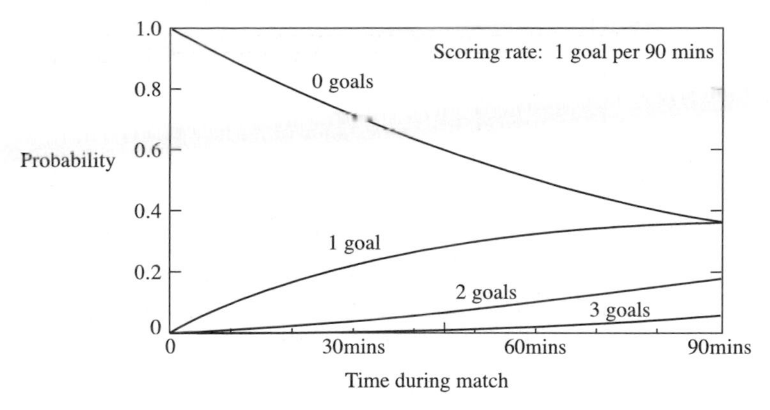

Figure 6.4. Probability of number of goals scored during a match for a team with an average scoring rate of one goal per 90 minutes.

probabilities given in the two graphs to calculate the probability of the various results.

For example, what is the probability that the stronger team wins 1–0? From the first graph the probability that the stronger team has scored 1 goal after 90 minutes is 0.33, and from the other graph the probability that the weaker team scores no goals is 0.37. The required probability is obtained by multiplying these separate probabilities together. So the probability that the result is 1–0 is $0.33 \times 0.37 = 0.123$.

The same procedure can be used to calculate the probability of any result and table 6.1 gives the probabilities for the 10 most likely scores. It also gives the probabilities expressed as a frequency. For example the 1–0 result has a probability of 0.123 or approximately 1/8, as this result would be expected in 1 in 8 such matches. The probability that the stronger team wins is obtained by adding the probabilities of all the scores for which this team wins including those not listed in table 6.1. This gives a probability of 0.49, just less than evens. The probability of a draw is 0.26 and of win for the weaker team is 0.25 – both about 1 in 4.

Clearly the scoring rates chosen for the above example were arbitrary and a similar calculation could be carried

Table 6.1

Score	Probability	Odds 1 in –	Result for stronger team
1–0	0.123	8	win
1–1	0.123	8	draw
2–0	0.092	11	win
2–1	0.092	11	win
0–0	0.082	12	draw
0–1	0.082	12	lose
1–2	0.062	16	lose
3–0	0.046	22	win
3–1	0.046	22	win
2–2	0.046	22	draw

out for any pair of rates. In fact it would be possible to make the model more sophisticated in many ways. For example, the scoring rate at any time could be allowed to depend on the score at that time as the teams adapt their strategies.

So far we have regarded the calculations as purely descriptive, but it is interesting that calculations of this sort can have implications for strategy. We shall now consider such a situation.

Strategy – a case study

In the previous chapter it was shown how, implicitly, the rules have been chosen to give a scoring rate which leaves the weaker team with a reasonable chance of winning. Looking at this from the point of view of teams in a match it is clear that a low scoring match benefits the weaker team and a high scoring match benefits the stronger team. This should, and no doubt does, affect the strategy of the teams. We shall examine this by considering matches between teams near the bottom and near the top of the Premiership.

Taking an average over four seasons the ratio of scoring rates in matches between teams finishing in the bottom five and the top five is approximately 3 to 7 so that, taking an average over these matches, the bottom teams score 3 goals while the top teams score 7. Assuming this ratio we can calculate the probability of each team winning the match. Putting this assumption another way, the probability that the weaker team will score the next goal is 0.3 and that the stronger team will score the next goal is 0.7. If only one goal is scored in the match the probability that the weaker team scored the goal, and hence won the match, is 0.3. The probability that the stronger team won is obviously 0.7.

Now consider a match with two goals. The only way to win the match is by scoring both goals. The probability of the weaker team scoring both goals and winning is $0.3 \times 0.3 = 0.09$ and the probability that the stronger team wins is $0.7 \times 0.7 = 0.49$. The probability of a draw is $1 - 0.09 - 0.49 = 0.42$. We see that the probability of the weaker team winning the two goal match is 0.09 compared with 0.30 for the one goal match, the probability of winning being reduced by a factor of more than three.

With higher numbers of goals the calculation is somewhat more complicated. For example with three goals there are four possible results: 3–0, 2–1, 1–2 and 0–3. Nevertheless the calculations are straightforward and table 6.2 gives the

Table 6.2

No. of goals	Probabilities		
	Weaker team wins	Draw	Stronger team wins
0	0	1	0
1	0.30	0	0.70
2	0.09	0.42	0.49
3	0.22	0	0.78
4	0.08	0.27	0.65
5	0.16	0	0.84
6	0.07	0.19	0.74

probabilities of the teams winning, losing and drawing for each number of goals in the match.

The pattern is rather complicated because of the possibility of draws with an even number of goals. However, the diminishing fortunes of the weaker team in higher scoring games is apparent. In games with an odd number of goals the chance of the weaker team winning decreases rapidly as the number of goals increases. With an even number of goals the probability of the weaker team winning is quite small although the decrease with the number of goals is slow. The compensatory probability of a draw falls rapidly. It seems that the defensive, low scoring, strategy adopted intuitively by weak teams playing stronger teams conforms to logic.

The basis of the scientific method is comparison of theory with the experimental facts. We can make such a comparison for the present theory by using results from the Premiership. Again we take matches between the teams finishing in the bottom five against teams finishing in the top five over four seasons. Figure 6.5 shows a comparison of the fraction of games won by the weaker teams with the theoretical

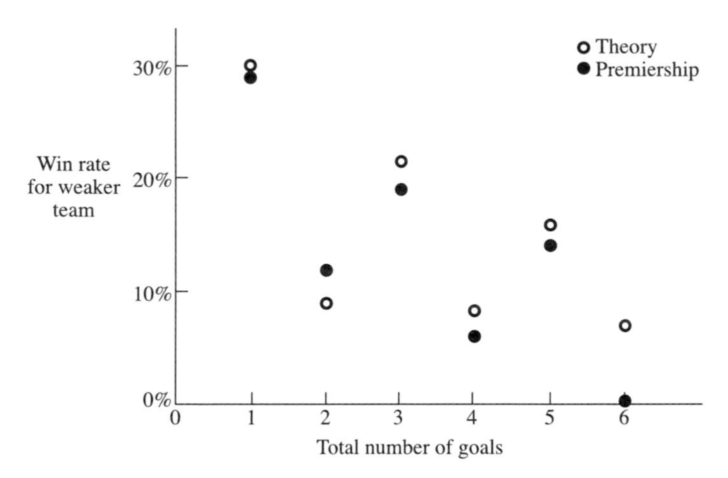

Figure 6.5. Dependence of fraction of games won by the weaker team on the number of goals in the match. Premiership results are compared with theory.

calculation. We see that, even though the model is a simple one, theory gives reasonable agreement with the results.

We need a goal!

It is a common situation that as the end of a match approaches it is essential to a team that they score a goal. For example, a team down 1–0 in a cup match needs a goal to take the match into extra time or to a replay. The strategy is clear – the team plays a more attacking game. In doing so its defence is weakened with an increased probability that their opponents will score. Can we give a quantitative description of these intuitive ideas?

We can define a team's chance of scoring in terms of a scoring rate, measured say in goals per hour. As our cup match approaches 90 minutes the losing team must increase its scoring rate and, for them unfortunately, increase their opponents' scoring rate also. Figure 6.6 shows the situation

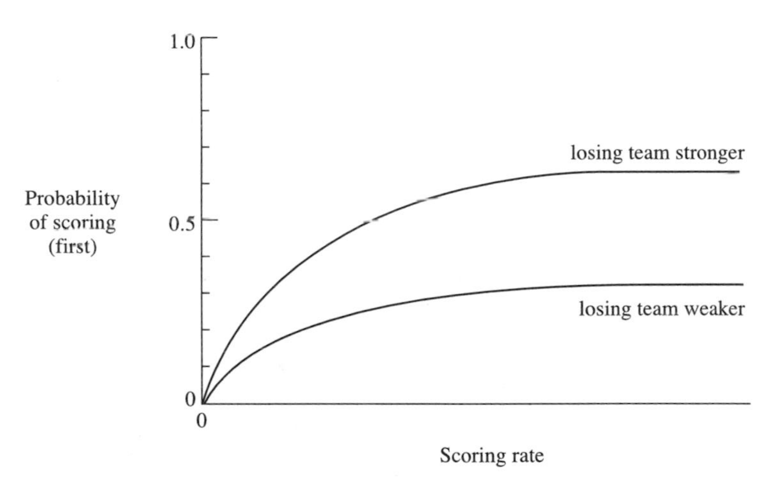

Figure 6.6. Dependence of probability of the losing team scoring in the remaining time for cases where the losing team has twice and half the scoring rate of their opponents.

at a given time, giving the probability of scoring the required goal in the remaining time without the opposition scoring. This clearly depends on the ratio of the scoring rates and the graphs given are for the cases where the losing team has a scoring rate of half, and twice, that of their opponents.

It is clear that the team must go all out for a high scoring rate and this is true independent of the quality of the opposition. However, while a very high scoring rate gives the team a probability of approaching 2/3 if they have twice the scoring rate of their opponent this is reduced to 1/3 when this ratio is a half. Nevertheless, the losing team must go for a higher scoring rate even when it makes it more likely that their opponents will score first.

A further insight can be obtained by recognising that the horizontal axis in figure 6.6 can be more completely defined as (scoring rate × time remaining). The consequence of this is illustrated in figure 6.7 for the case of equal scoring rates. The graph illustrates how, no matter what the scoring rate, the probability of scoring the required goal remorselessly approaches zero as time runs out.

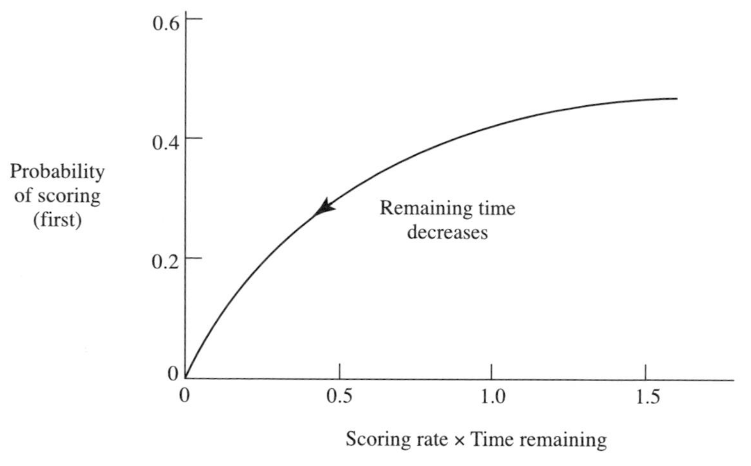

Figure 6.7. Probability of scoring plotted against the product scoring rate × time remaining, for teams of equal scoring rates.

The off-side barrier

The need for an off-side rule has been accepted from the earliest days. Indeed the first off-side rule was more stringent, requiring that there be three opposing players in front of an attacking player when a pass is made, rather than the present two. The rule has a crucial influence on the way the game is played. Without it, attacking players could congregate around the goal to receive long passes from their colleagues, as happens at corner kicks.

Essentially the rule allows the defenders to create a barrier beyond which the attackers cannot stray. The barrier can be broken by an attacking player either by his taking the ball past the defenders, or by a well-timed run. To achieve a well-timed run the attacker must either react more quickly than the defenders to a pass aimed behind their line, or he must anticipate the pass and be running at the time it is made.

The most efficient way of thwarting the defence is for a colleague to kick his pass when the attacker is already moving at full speed past the last defender. The maximum advantage is gained if the defenders only react at the time of the pass. Figure 6.8 illustrates the movement of the attacker

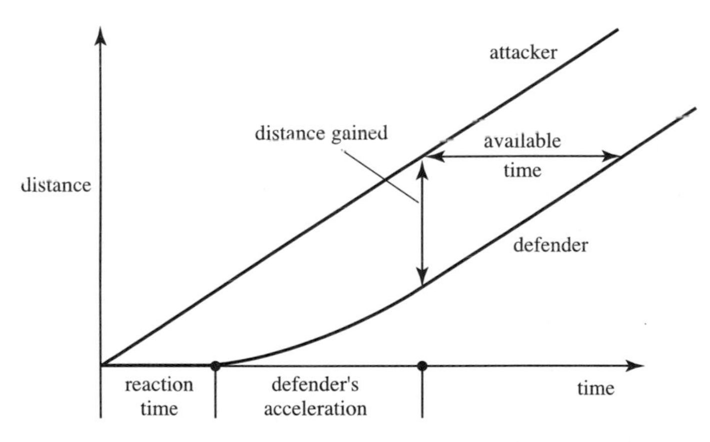

Figure 6.8. Diagram showing the movement of an attacker attempting to defeat the off-side barrier and the response of a defender.

and a defender during this tactic. The attacker has the advantage, firstly of the defender's reaction time, and secondly of the defender's need to accelerate. Typically each of these factors gives the attacker half a second and if he is running at, say, 12 miles per hour, this means that he would be clear of the defender by 6 yards. The figure shows the time this makes available to the attacker to make his next move, free of the defenders' attention. Whether he can fully exploit this will, of course, depend on the quality of the pass and his ability to bring the ball quickly under control.

Intercepting a pass

When the ball is passed along the ground to a colleague care is taken to avoid the pass being intercepted. Conversely, opposing players look for an opportunity of preventing a successful pass. What is the requirement for a successful interception? There are three situations to consider as illustrated in figure 6.9.

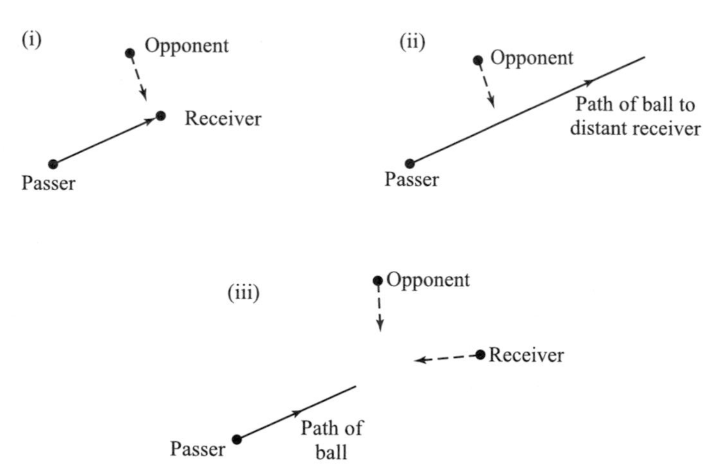

Figure 6.9. (i) Opponent too distant to intercept. (ii) Receiver too distant to intervene, opponent may or may not be able to intercept. (iii) Both players can run for the ball.

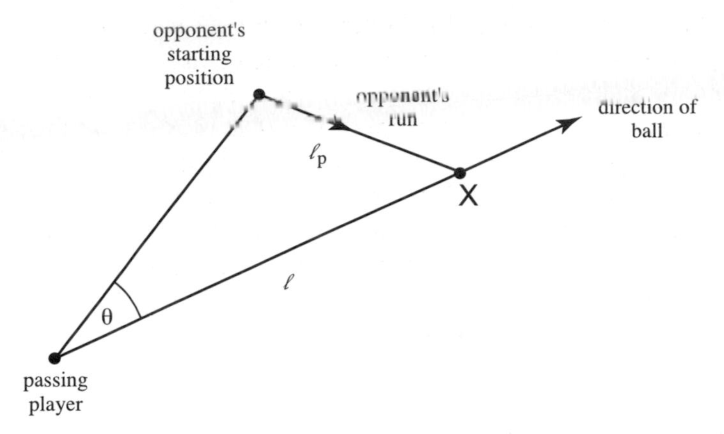

Figure 6.10. Diagram showing the positions of the passing player and an opponent together with the paths of the ball and the opponent's intercepting run.

The first case is the simple one of the short pass where the receiving player is sufficiently close to the passing player that the nearest opponent cannot intervene. The second case is that of the long pass where the receiving player is so distant that he cannot affect the outcome by moving toward the ball. The question then is whether the opponent can intercept the ball on its path to the receiver. In the third, more complex, case the movements of both the receiver and the opponent are involved.

In the second case the ball is passed at an angle θ to the line joining the passing player and the potentially intercepting opponent as shown in figure 6.10. For an interception there must be a point along the ball's path which the opponent can reach in less time than that taken by the ball. If the ball travels with a speed s_b the time taken for it reach the point X, a distance ℓ from the passer, is ℓ/s_b. The time taken for the opponent to reach X at speed s_p is ℓ_p/s_p. From the geometry these times can both be calculated.

Figure 6.11 gives the result of such a calculation for the case where the player runs at half the speed of the ball. The first part of the figure plots the time taken for the ball and the opponent to reach the distance ℓ along the ball's path

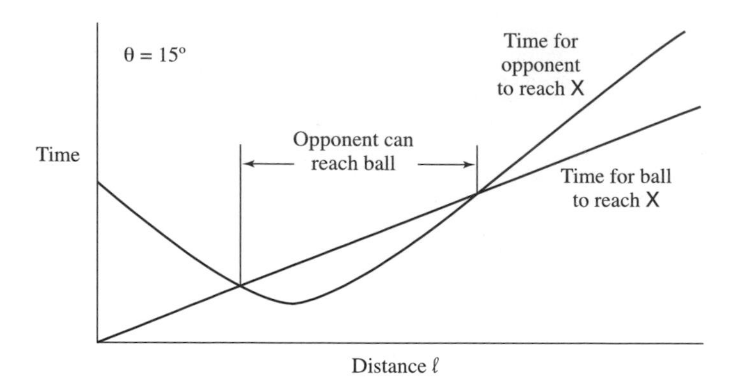

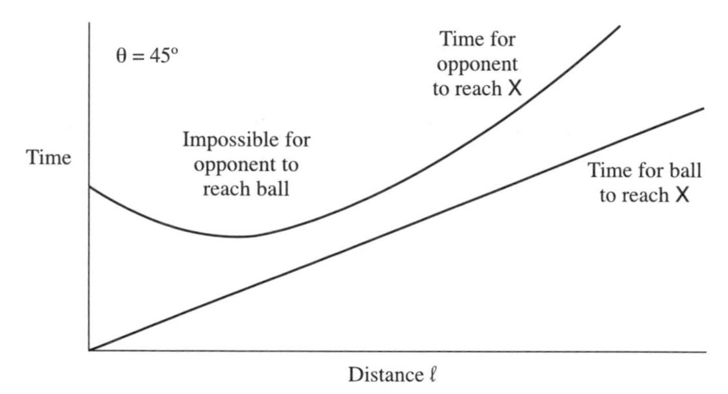

Figure 6.11. Times for the ball and the opponent to reach X over the range of distances, ℓ. For the $\theta = 15^\circ$ case the lines cross and interception is possible. For $\theta = 45^\circ$ no interception is possible. In this example the speed of the ball is twice that of the opponent.

for an angle $\theta = 15^\circ$. It is seen that, provided the receiving player is at too great a distance to intervene, there is a band of ℓ where the opposing player can reach a point X before the ball, and can therefore successfully intercept it. The second part of the figure plots the same quantities for a more conservative pass with $\theta = 45^\circ$. In this case it is not possible for the opponent to intercept the pass no matter which direction he takes.

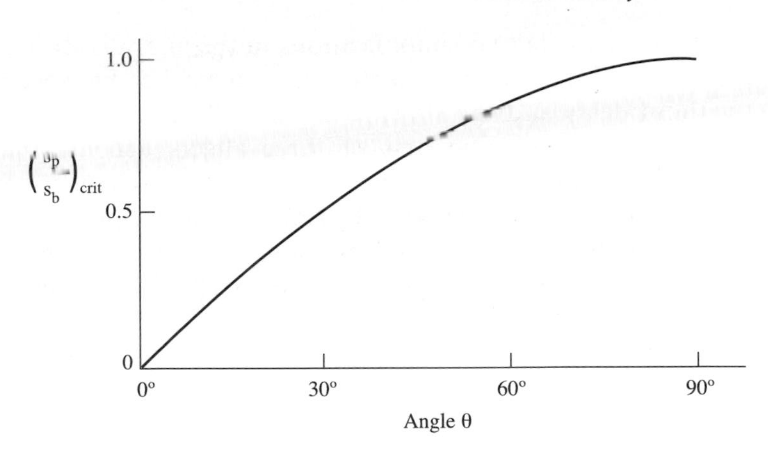

Figure 6.12. Graph of the critical ratio s_p/s_b against the angle θ. Interception is possible for ratios above the curve.

It turns out that for any angle θ of the pass there is a critical ratio of the speed of the player to that of the ball which must be exceeded if a successful interception is to be made. Figure 6.12 gives a graph of the critical ratio of s_p/s_b against θ.

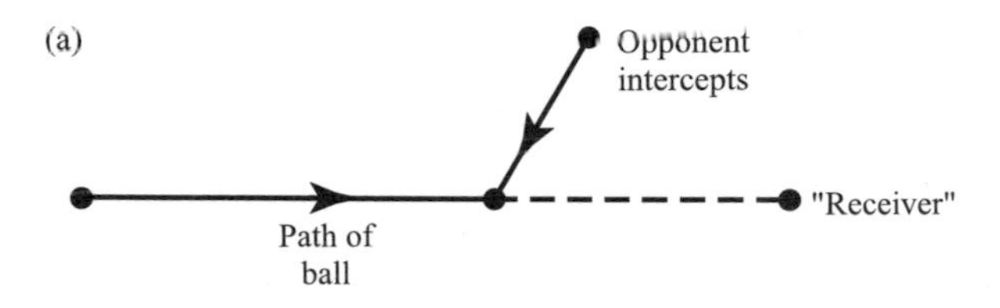

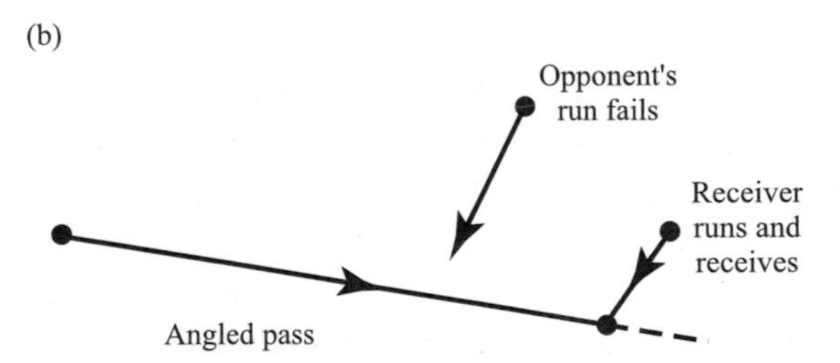

Figure 6.13. The direct pass in (a) would be intercepted whereas an angled pass as in (b) would be successful.

We have made two simplifications in the analysis. It has been assumed that the intercepting player reaches his speed s_p without delay and the slowing of the ball during the pass has been neglected. The first of these effects benefits the passer of the ball and the second benefits the opponent.

The third case, where both the receiver and the opponent move to the ball, includes the situation where a pass aimed directly to the receiving player can be intercepted as in figure 6.13(a), whereas an angled pass would be successful as in (b).

Chapter 7

	FOOTBALL LEAGUE 1888–89	P	W	D	L	F	A	Pts
1	Preston	22	18	4	0	74	15	40
2	Aston Villa	22	12	5	5	61	43	29
3	Wolves	22	12	4	6	50	37	28
4	Blackburn	22	10	6	6	66	45	26
5	Bolton	22	10	2	10	63	59	22
6	WBA	22	10	2	10	40	46	22
7	Accrington	22	6	8	8	48	48	20
8	Everton	22	9	2	11	35	46	20
9	Burnley	22	7	3	12	42	62	17
10	Derby	22	7	2	13	41	60	16
11	Notts County	22	5	2	15	39	73	12
12	Stoke	22	4	4	14	26	51	12

7

The best team

Football clubs covet trophies as a symbol of their success. For many clubs the most satisfying achievement is to win their league championship. This is certainly true in the Premiership where the strongest teams proclaim the importance of the Championship as compared with the winning of the F.A. Cup. If a team wins the Championship they have demonstrated that they are the best team in England. Or have they?

If the Championship is won by a single point, then it is possible to reflect on the occasions during the season where a point was won through a lucky shot, a goalkeeping error or a wrong decision by a linesman or referee. The team that came second could just as well have won the Championship.

On the other hand, if the winning team finishes well ahead of its competitors we feel more confident that it has shown itself to be the best team. Can we quantify this subjective assessment to obtain a probability that the winning team is the best team?

A thought experiment

Let us start by imagining a league in which all of the teams are equally good. For simplicity let us first assume that each match is equally likely to be won by each contestant. What will the final league table look like? It is obvious that the teams will not all obtain the same number of points. There will be a 'champion team' (or teams) and there will be a spread of points throughout the league determined entirely by chance.

To make our 'thought experiment' more precise we shall allocate probabilities to each type of result. The concept of probability was introduced in chapter 5 and we recall that mathematical probability is measured on a scale of 0 to 1, a probability of 1 corresponding to certainty, and a probability of 0 to no chance. For example, with a thrown dice the probability of each number is 1/6, the sum of their probabilities being 1 as we would expect. The probability of an even number being thrown is 1/2. The probability 1/2 can also be described as 50% (50/100) and we shall sometimes use the percentage terminology for convenience.

Returning to our experiment we allocate 1 point to each team for a draw. In professional matches the frequency of drawn games is close to one in four, and so in our model we shall take the probability of a draw to be 1/4. The probability that the match is won is therefore $1 - \frac{1}{4} = \frac{3}{4}$, and since the teams are equal they both have a 3/8 chance of winning. A winning team takes 3 points and a losing team none. This gives us the probability table for each match (table 7.1).

We can now 'play' a season's matches with these probabilities. This is easily done using a computer or a calculator to provide random numbers. A 'league table' from such a calculation is given in table 7.2. Our league has 20 teams who play each other twice.

Table 7.1

Result	Points	Probability
Win	3	$\frac{3}{8}$ (37.5%)
Draw	1	$\frac{1}{4}$ (25%)
Lose	0	$\frac{3}{8}$ (37.5%)

Table 7.2

	W	D	L	Points
1	19	10	9	67
2	18	9	11	63
3	18	8	12	62
4	17	10	11	61
5	16	10	12	58
6	16	8	14	56
7	13	16	9	55
8	15	9	14	54
9	16	5	17	53
9	15	8	15	53
9	15	8	15	53
9	14	11	13	53
13	13	13	12	52
14	14	9	15	51
14	13	12	13	51
16	15	4	19	49
17	11	11	16	44
18	9	16	13	43
19	9	8	21	35
20	8	7	23	31

We see that there is a clear champion with 67 points and that the spread between the top and bottom teams is 36 points – all this with precisely equal teams. In the Premiership the champion teams obtain an average of about 80 points and the spread from top to bottom is about 50 points. It is clear, therefore, as we would expect, that the spread of abilities of the real competing teams adds to the spread of points. It is also clear, however, that randomness makes a large contribution.

A better team

Before looking at the question of whether the champion team is the best team let us carry out one more computer simulation.

Table 7.3

Result	Points	Probability for the better team	Probability for the rest (against each other)
Win	3	$\frac{9}{20}$ (45%)	$\frac{3}{8}$ (37.5%)
Draw	1	$\frac{1}{4}$ (25%)	$\frac{1}{4}$ (25%)
Lose	0	$\frac{6}{20}$ (30%)	$\frac{3}{8}$ (37.5%)

We will add to the egalitarian league of the previous simulation one team which is better than the rest. We shall still give it a probability of a quarter for a draw for its games against the other teams, but make it more likely to win than lose the remaining games with probabilities in the ratio 3 to 2. Thus the probability is as shown in table 7.3.

The better team is now allowed to play the rest and the results are included with the previous ones to compile a new league table as shown in table 7.4.

With the allocated probabilities the average number of points expected for the better team from 40 matches is

$$40 \times \left[\left(\frac{9}{20} \times 3\right) + \left(\frac{1}{4} \times 1\right)\right] = 64 \text{ points.}$$

In the simulation the team actually did better than this, scoring 67 points. Nevertheless it only came second. A less able but more lucky team scored 71 points. Of course, other simulations using the same probabilities would give different results, and sometimes the best team would be 'champion'. However, for the given probabilities it can be shown mathematically that most times the better team will not come out on top.

We see, therefore, that even without a difference of ability there is a spread in the distribution of points, and that with a difference in ability a team with greater ability than the rest is not guaranteed top place.

In the simulations described above the probabilities were given and the distribution of points was calculated. We now

Table 7.4

	W	D	L	Points	
1	20	11	9	71	
2	17	16	7	67	Better team
3	18	11	11	65	
4	17	12	11	63	
5	18	8	14	62	
6	17	9	14	60	
7	16	10	14	58	
8	16	9	15	57	
9	15	11	14	56	
10	14	13	13	55	
10	13	16	11	55	
12	16	6	18	54	
12	15	9	16	54	
14	16	5	19	53	
14	15	8	17	53	
16	14	11	15	53	
17	13	12	15	51	
18	11	13	16	46	
18	10	16	14	46	
20	9	10	21	37	
21	8	8	24	32	

come to the more realistic but more difficult problem where, at the end of the season, the distribution of points is given and we would like to know the probability that the champion team is the best team. However, before analysing this problem we examine two general features of probability theory.

Concerning probability

Our assessment of probability depends on the information available. For instance, let us ask the probability that a randomly chosen Premiership match was drawn. Since about a quarter of such matches are drawn the answer is approximately 25%. If we are then told that one of the teams scored more than one goal, a draw is less likely, the probability being reduced to about 5%. If we are told that the total number of goals in the match is odd, the probability of a draw is zero. We see that information alters probability.

Another situation arises when we want to extract information from a sample of data. The larger the sample the more confident we can be about our conclusions. Imagine, for example, that we are supplied with a team's results for a particular completed season and that they are given one at a time. With a few results we obtain only a hint as to how many points the team obtained that season. As the number of results supplied increases the probable outcome becomes clearer, and finally becomes certain when all of the results have been given. It is clear that increasing the size of the database improves our assessment of probability.

The best team in the Premiership

We now turn to the problem of deciding the probability that the team winning the Premiership is the best team. Clearly the top team is the most likely to be the best team, but can we put a probability to it? There is no limit to how sophisticated our method could be, but we will aim for the simplest procedure which satisfies some basic requirements.

First, it should say that if two teams finish equal top, they are equally likely to be the best team. Next, the probability of the top team being best should increase with increased points difference over the rest of the teams. If the top team has a few points more than the runner-up it is more likely to be the best team than with only a one point difference. Finally, with a very large points difference the probability that the top team is the best must approach 100%.

We will measure a team's quality by its 'points ability'. We define this as the number of points it would have obtained if the random effects had averaged out, there then being no advantage or disadvantage from these effects. The most

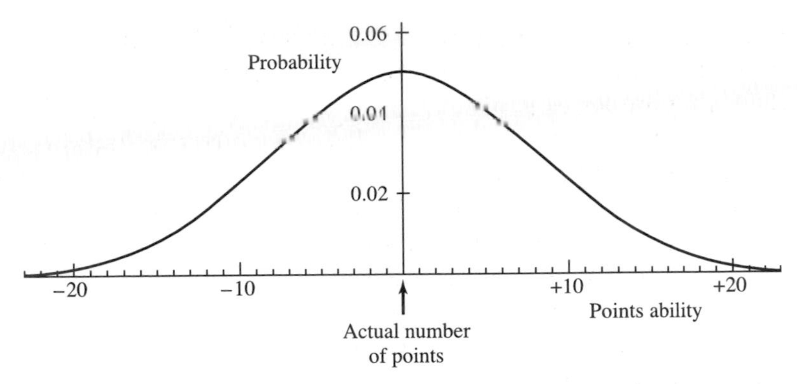

Figure 7.1. Smoothed graph giving the probability (per point interval) of a team's points ability differing from the actual number of points achieved.

likely value of a team's points ability is the number of points it actually achieved, but because of random effects there will be spread of possible values. We shall take the probability of a given points ability to have the bell-shaped form shown in figure 7.1. Technically this is called a normal distribution. For simplicity we take the spread in possible points ability to be given by the spread which purely random results would give. It is seen from the graph that the most probable points ability is the actual number of points gained, the probability being 0.05 (5%). For a difference of 8 points the probability has fallen to 3% and for a difference of 16 points to less than 1%.

The calculation required is quite subtle. We must consider *all* possible values of the top team's points ability and for *each one* we must take account of *all* the possible points abilities of *all* the competing teams. We shall illustrate the procedure by taking an example. For a chosen value of the top team's conjectured points ability we shall first determine the probability that the runner-up has a lower points ability. This then has to be repeated for all possible values of the top team's points ability and the probabilities for each case then added to give the probability that the top team is better than the runner-up. This example will illustrate the procedure.

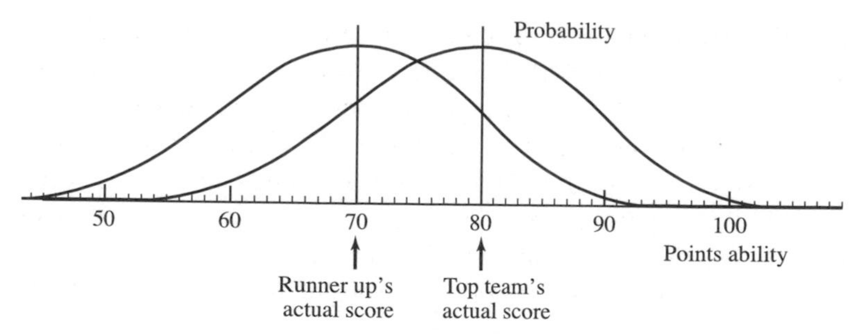

Figure 7.2. Probability curves for the top team and the runner-up for a case where their actual points difference is 10.

The actual calculation allows for all contenders, not just the runner-up.

In our example we take a case where the top team has achieved 80 points and the runner-up 70. To determine the likelihood that the top team has a higher points ability than the runner-up we need the bell-shaped curves for both, and these are shown in figure 7.2. Again for example, we first take the points ability of the top team to be lower by 4 than the points actually obtained, as shown in figure 7.3. The probability of this is measured by the height, p_1, of the curve at this point, which is 0.044. The top team will then be a better team than its rival if the rival's points ability is lower still. This is illustrated in figure 7.3, where the range of the rival's points

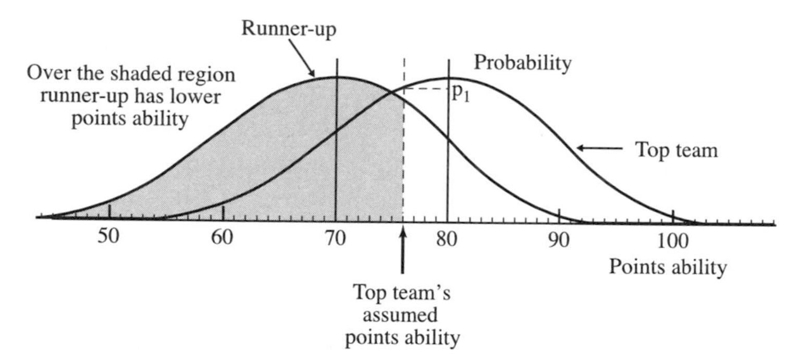

Figure 7.3. Illustrating the calculation for the case where the top team's points ability is 4 points below its actual number of points.

abilities over which it is less than the top team's is shown shaded. The probability that the runner-up's points ability lies in the shaded region is the sum, p_2, of the probabilities in this region, in the present example 0.77. The combined probability that the top team's points ability has the chosen value and that the runner-up's points ability is less is the product p_1p_2, which here is $0.044 \times 0.77 = 0.034$ or 3.4%.

But this was just for the example of the points ability of the top team being 4 points lower than the points it actually obtained, a points difference of −4. We must now take all possible points differences ... −4, −3, −2, −1, 0, +1, +2, +3, +4 ... and repeat the calculation for each. The total probability that the top team is the better is then the sum over all these cases. In the present example, with a points difference over its rival of 10 points, the probability that the top team is the better team is 81%. Correspondingly the probability that the rival team is actually the better team is 19%.

We now have to recognise that for a team to be the best it is not sufficient just that it be better than its closest rival. It must be better than all of the other teams. The required calculation is similar in principle to that described above but is a little more complicated. For each value of the top team's possible points ability it is necessary to calculate the probability that all the other teams have a lower ability. These probabilities are then summed to obtain the probability that the top team is the best. This calculation can be repeated for any other team to determine the probability that, although it didn't come top, it is the best team. Using this procedure we can carry out the calculation for any season's results. Let us first look at the first season of the Premiership, 1992–93.

The first Premiership season

In the first Premiership season Manchester United won the Championship and were 10 points clear of the second team, Aston Villa. Aston Villa were followed closely by Norwich

City and Blackburn, there then being a large gap down to the next team Queen's Park Rangers. This means that we only have to consider the top four teams. Their part of the points table is given below.

	Points
Manchester United	84
Aston Villa	74
Norwich City	72
Blackburn	71

The calculation gives Manchester United a probability of being the best team of 68%. The table of probabilities for the four clubs is

	Probability that team is the best team
Manchester United	68%
Aston Villa	14%
Norwich City	10%
Blackburn	8%

Manchester United have a five times higher probability of being the best team than Aston Villa.

It might seem that with a 10 point lead the probability that Manchester United be the best team should be more than 68%. However, such a judgement is probably influenced by the prestige associated with the team actually being Champions. It perhaps makes the level of uncertainty implied by 68% more plausible when we note that of Manchester United's 42 matches, the result of 28 could have been changed by a single goal. This gives some insight into the role of chance in determining the number of points obtained. The other three teams involved all had a similar number of results decided by one goal, further indicating the part randomness plays in determining the outcome.

Other years

In the first nine years of the Premiership the competition was won seven times by Manchester United. The Champions in the other two years were Blackburn and Arsenal. In both cases these teams were only one point clear of Manchester United. It is not surprising therefore that, allowing for all the other teams involved, the probability that Blackburn and Arsenal were the best teams in the Championship in their winning years was less than 50%, being 48% for Blackburn and 49% for Arsenal.

Manchester United's best season was 1999–2000 when they were 18 points ahead of their rivals, with a 92% probability that they were the best team. Our judgement of these figures for each year is very likely affected by the fact that we are aware of the results over several years. The analysis can be extended to cover any number of years and as an example we can look at the first five years of the Premiership. The result, which coincides with our intuition, is that the probability that Manchester United were the best team over this period is 99.99%.

The difference between this figure, which corresponds almost to certainty, and the results for the individual seasons might be a little surprising. It is explained by the factors mentioned in the earlier discussion of probability. Firstly, that our assessment of probability depends on the information available and, secondly, that a larger sample allows greater confidence.

Another view

Some readers might find the distinction between the 'best team' and the team which wins the Championship difficult to accept. That is quite reasonable since the concepts involved are rather theoretical and the assumptions made for the purpose of simplicity were not treated rigorously.

An alternative view of the calculations is that they provide a figure-of-merit which enables us to rank champions according to their superiority over all the other teams. This provides a more sophisticated measure than just taking their points lead over the runner-up. Seen as a figure-of-merit the results of the calculations fit quite well our intuitive assessments. Clearly Manchester United's performance in their record season 1999–2000 with a figure of merit of 0.92 was better than in its first Premiership Championship with 0.68 and was certainly better than Blackburn and Arsenal's narrow wins for which the figures-of-merit were 0.48 and 0.49.

The Cup

It is regarded as a special event when a team wins 'the double' – the League Championship and the F.A. Cup. This happened only seven times in the years from 1946 to 2001. Since we have been involved with probabilities in this chapter it is perhaps appropriate to analyse the performance of the Champion teams to see why they have a low success rate in the Cup.

Looking at the statistics since 1946 the team destined to win the Championship has a better than 50/50 chance of winning in each round of the Cup, including the Final. In the first four rounds in which they play (third round to quarter final) they are three-to-one favourites to win in each round (before the draw is made). In terms of probabilities the probability that they will win through the round is 3/4.

Using this figure we can calculate the probability that they will win through all the first four rounds. This is obtained by multiplying together the probabilities for winning each round. So the probability is $\frac{3}{4} \times \frac{3}{4} \times \frac{3}{4} \times \frac{3}{4} = 0.32$, which is close to 1/3, giving them only a one-in-three chance of reaching the semi-final.

A top team playing in a semi-final or final match has a 5/8 chance of winning and so the probability of their winning both matches is $\frac{5}{8} \times \frac{5}{8} = 0.39$. We can now calculate the probability that the team due to win the Championship will also win the Cup. To do this it must win through the first four rounds with a probability 0.32 and then win the semi-final and final with a probability of 0.39. The overall probability is therefore $0.32 \times 0.39 = 0.125 = \frac{1}{8}$.

So the chance of the team which wins the League or Premiership also winning the Cup is one-in-eight. For the 56 seasons from 1946 this predicts seven double wins which, as mentioned earlier, is the actual number.

Chapter 8

8

The players

Footballers with outstanding ability are usually recognised while still at school. Those who succeed and play at the highest level are either identified and chosen by a top club at an early age or have demonstrated their ability playing at a lower level.

Many players showing early potential only have brief stays in the professional game, but the most successful players have professional careers lasting about 15 years, typically between the ages of 20 and 35. Most players reach their peak of ability in their middle 20s. Once past 30 it becomes increasingly difficult to hold a place at the top. This is illustrated by the graph in figure 8.1 which gives the number of players at each age in the Premiership. The graph has been smoothed to remove statistical variations.

A more selective measure of the peaking in ability of the best players is the readiness of clubs to pay a high transfer fee. Figure 8.2 shows a graph of the percentage of transfer fees of over a million pounds taking place at each age. It is seen to be more sharply peaked than the first graph, its maximum occurring at the age of 26 as compared with 22. This is partly due to the fact that clubs are buying proven players. On the other hand, the clubs are investing in the future of the players, some of whom will not have reached their peak.

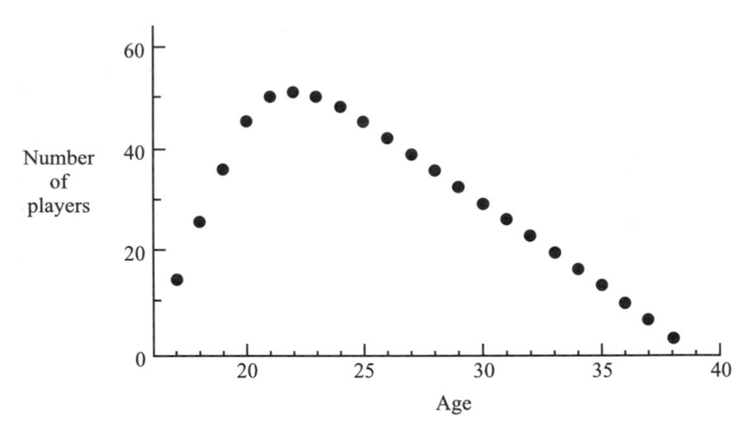

Figure 8.1. Number of players of each age in the Premiership.

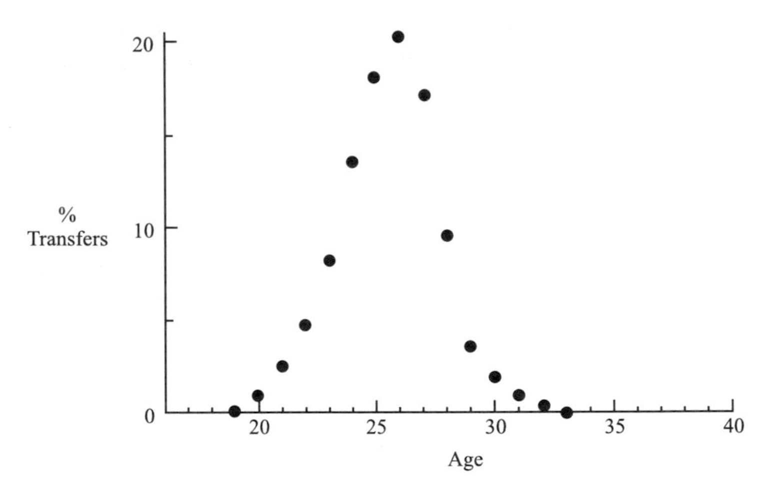

Figure 8.2. Percentage of transfers in excess of a million pounds at each age in the Premiership.

A remarkable statistic

In analysing the age structure of the profession it becomes apparent that, in addition to the dependence on age, there is a dependence on birth date. Figure 8.3 shows the percentage of players in the Premiership born in each month of the

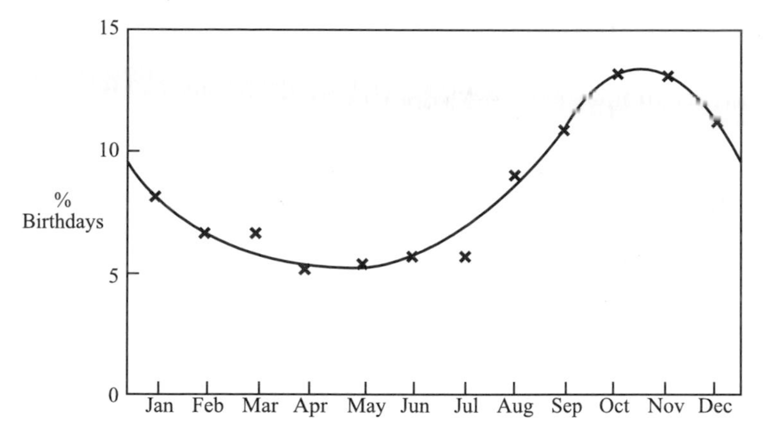

Figure 8.3. Percentage of Premiership players born in each month of the year.

year. The amazing result is that the probability of reaching this level is more than twice as high for boys born in the autumn as for those born in the summer.

The likely explanation seems to be that the intake to each school year is defined by the child's age around the summer holidays. This means that those born in the autumn will be the best part of a year older than those born in the summer. On average, therefore, they will be slightly taller and stronger, and the effect of an almost one year difference will be particularly important at an early age. Consequently those born in the autumn will have a better chance of being selected for the school team. This advantage is then amplified by the practice which results from playing in the team. Presumably the cumulative effect of this process throughout their school careers leads to their higher level of success.

It seems unlikely that innate ability depends on birth-date, and perhaps professional clubs could gain some advantage by making an allowance for this factor in identifying prospective players.

It will no doubt occur to the reader that the distribution of birth-dates in the general population might also show a seasonal bias. In fact the birth rate has only a small variation throughout the year and is highest in the summer.

Careers

Almost all boys have the opportunity to play football at some time and those with aptitude or enthusiasm will play for their school or local team. It seems likely that many, if not most, of the youngsters would accept an offer of a place in professional football. This means that the market is very competitive. Something like one in a thousand boys will play at some time in one of the top four professional leagues, nowadays the Premiership plus Divisions 1 to 3.

Most professionals spend their careers in the lower leagues and only one in a hundred English professionals will play for the England team. Many players who reach the professional ranks have rather brief stays and the average professional career is about six years. Figure 8.4 gives a smoothed graph of the percentage of players who have careers of a given length in the top four leagues, and the percentage whose careers exceed a given length. We see from the first graph that almost a quarter of the players spend only one season in the top leagues. The second graph shows that most players stay in the top leagues for less than five years.

It is not surprising that the better players have a longer career, sometimes extending it by taking an Indian summer in the lower leagues. The best players typically play professional football for about 20 years. The record is held by Stanley Matthews who played until he was 50 years old and had a playing career lasting 33 years.

Heights of players

One of the merits of football is that players of all sizes can enjoy the game and succeed at the highest level. This gives soccer an advantage over many other games in which height or weight are crucial.

Nevertheless, height can have an influence in deciding the role which best suits each player. The clearest example is

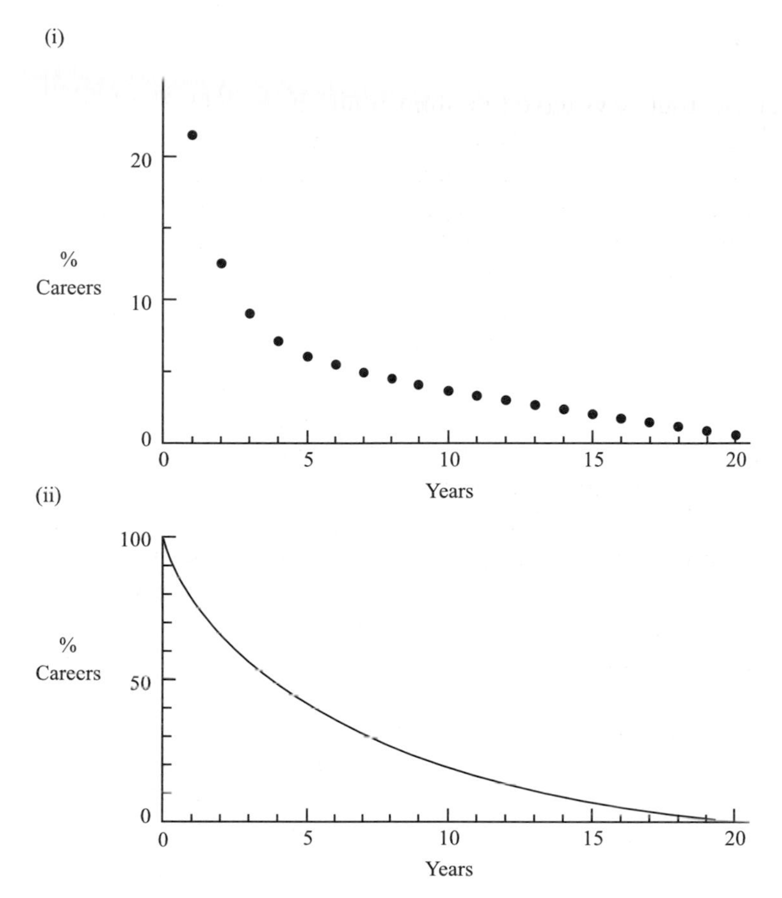

Figure 8.4. Graph of (i) the percentage of careers against the career duration and (ii) the percentage of careers exceeding a given number of years.

that of goalkeepers. There is obviously an advantage in being tall because of the need to deal with high shots and with balls crossed into the goal area. This is reflected in the heights of successful goalkeepers. To illustrate this figure 8.5 compares the distribution of heights of young men generally with those of goalkeepers, defenders and forwards in the Premiership. It is seen that it is rare for a goalkeeper to be under 5′ 10″ and that the most common height is about 6′ 2″,

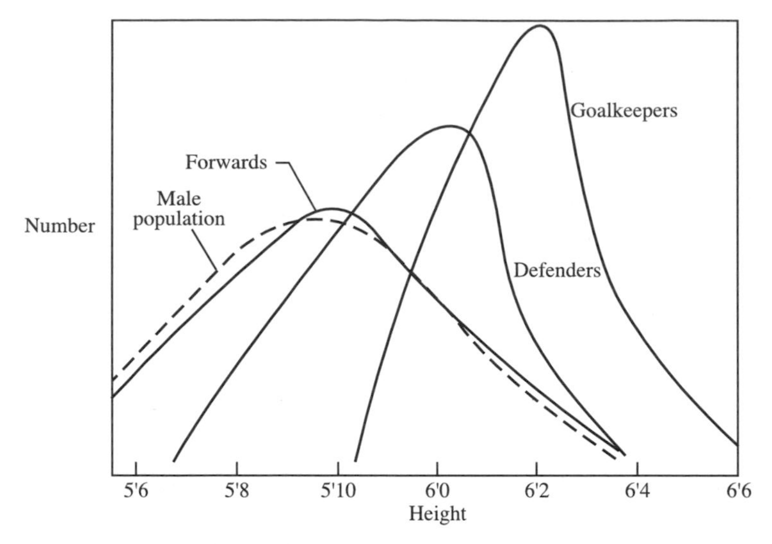

Figure 8.5. Distribution of heights for goalkeepers, defenders and forwards compared with the general adult male population of similar age.

several inches above the average height of the general male population.

Although less pronounced than for goalkeepers there is a tendency for defenders to be above average height. This presumably arises from the need to compete to head high balls. Forwards are seen to have a height distribution close to that of the general population with a peak at about $5' \, 10''$.

Strikers

Strikers receive much of the glory in football matches but are vulnerable to the constant attention given to their scoring performance, which is readily measured. Figure 8.6 gives a graph of the average scoring rate for professional strikers plotted against age. It is seen that they typically reach their peak around the age of 23. It is rare for strikers to carry a high scoring rate into their thirties, John Aldridge being a remarkable example of one who did.

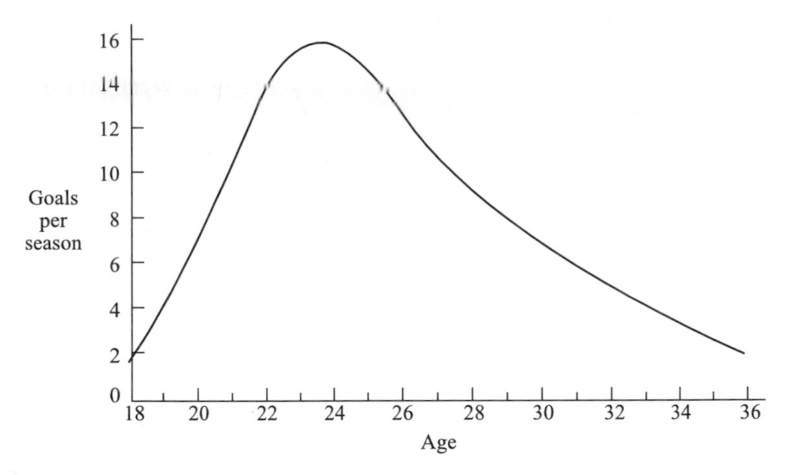

Figure 8.6. Smoothed graph of goals scored per season by strikers at each age.

The number of times a player is selected to play for his country gives some measure of his success. Apart from this there is no quantitative measure which is generally applicable. For strikers goal-scoring provides such a measure. However, this is not straightforward because the number of goals scored depends upon the degree of opportunity. Let us look at the elite among England's strikers.

The simplest measure for international strikers is the total number of goals scored. This is given in table 8.1 for the top

Table 8.1. Top England goalscorers

	Goals
Charlton	49
Lineker	48
Greaves	44
Finney	30
Lofthouse	30
Shearer	30
Platt	27
Robson	26
Hurst	24
Mortensen	23

Table 8.2

	Matches	Goals ×	Scoring rate =	Figure of merit
Greaves	57	44	0.77	34
Lineker	80	48	0.60	29
Lofthouse	33	30	0.91	27
Charlton	106	49	0.46	23
Mortensen	25	23	0.92	21
Lawton	23	22	0.96	21

ten scorers among those who have played since 1945. Charlton and Lineker appear at the top with Greaves not far behind. But this table does not allow for the number of games played. Lawton, for example played only 23 games, but scored 20 goals.

We cannot take the scoring rate, that is goals per game, as a measure because, for example, a player who played once and scored two goals would go above all of the players in our list. A proper measure calls for a 'figure of merit'. Unfortunately figures of merit are bound to be subjective. Nevertheless, let us look at a figure of merit which gives equal weight to the total number of goals scored and to the scoring rate. This is obtained by multiplying the two together. Table 8.2 shows the result; each person can judge whether this procedure has, for them, caught the essence of success for goalscorers.

Taking a longer perspective, Steve Bloomer (1895–1907) with 28 goals in 23 matches also has a figure of merit of 34, and George Camsell (1929–36) who averaged two goals per match over 9 matches has a figure of merit of 36.

Composition of teams

The composition of teams has attracted a lot of interest in recent years, mainly due to the large influx of foreign players attracted by the large salaries which the Premiership can offer. An extreme example was the Chelsea team which won the F.A. Cup in the year 2000. The team fielded had only one British player, Wise. This can be compared with the Chelsea team which won the Cup in 1970. That team was entirely British and five of the players were born in London.

When football started in the late nineteenth century the players in each team were drawn from the same school or the same locality, so the players had that in common with each other and also with their supporters. It is easy to understand why people would support a team if they know the players, or at least could feel that the team represented the local community. While this situation persists at the lower levels of football it has long since been transformed in the professional game.

The final of the first F.A. Cup competition after the second world war was played in 1946. The winning team was Derby County. That team had only three players born in Derbyshire. Since then teams have typically had two or three local players but there has of course been some variation. When Everton won the Cup in 1966 they had five Merseysiders in their team but Liverpool, winners in 1986, had no English players at all.

It is perhaps surprising that the pattern of mainly non-local players goes back a hundred years. For example, at the end of the nineteenth century the Leicester team, then Leicester Fosse, typically had two players born in the county. This has remained roughly the same for a hundred years. It is interesting to note that throughout the twentieth century the Leicester team usually had as many Scots as Leicester born players.

No-one would have predicted the modern developments or the remarkable fact that most football fans give their continuous support to teams which in almost no way represent them. Youngsters often confer their allegiance on teams they have never seen, and remain loyal thereafter. The whole business is mysterious but, without a doubt, club loyalty is a crucial part of the modern game and provides much excitement for the fans.

Although any player can be eligible to play for any club the situation for international players is of course quite different. To play for a national team the normal qualification is that you were born in the country. For some countries the national identity is diluted by players whose qualification comes from having a parent born in the country. Almost all the players who play for England were born there.

The continuity of the players' allegiance to their country gives a continuity to the national team which is largely absent from professional club teams. The composition of the national team changes slowly as young players develop and replace the older stalwarts.

Players' origins

A simple investigation of the origins of top players can be made by looking at the birthplaces of most successful members of England teams. The list below gives the birth places of England players who have played more than 60 times for England since 1945, and the locations are shown on the map of England (figure 8.7). It is seen that there is a general correlation with the centres of large populations, with London, the Midlands and the North being well represented. It would be interesting to carry out a statistical analysis, allowing for population levels, to find out which places contribute more than their share of top players.

T. Adams	Romford
A. Ball	Greater Manchester
G. Banks	Sheffield
J. Barnes	Jamaica
T. Butcher	Singapore
R. Charlton	Ashington, Northumberland
R. Clemence	Skegness
T. Finney	Preston
E. Hughes	Barrow

K. Keegan	Doncaster
G. Lineker	Leicester
R. Moore	Barking
S. Pearce	Hammersmith
M. Peters	Plaistow, London
D. Platt	Oldham
B. Robson	Chester-le-Street, Durham
K. Sansom	Camberwell
D. Seaman	Rotherham
A. Shearer	Newcastle
P. Shilton	Leicester
C. Waddle	Newcastle
D. Watson	Stapleford, Nottinghamshire
R. Wilkins	Hillingdon
R. Wilson	Shirebrook, Derbyshire
W. Wright	Ironbridge, Shropshire

Figure 8.7. Map showing the birthplaces of top England players.

Table 8.3

Footballer of the Year Awards		
	Number of awards	Awards per million of population
England	29	0.062
Wales	2	0.072
Scotland	9	0.172
N. Ireland	4	0.262

Historically many of the great players in the English league have come from the other countries of the United Kingdom – Scotland, Wales and Northern Ireland. The reason for this arises from the comparatively large population of England, comprising over 80% of the UK's population. This means that the large and wealthy clubs are predominantly in England, and players in the smaller countries are then attracted to these clubs by the higher wages.

However, there is more to be explained. We can assess the contributions from different countries by analysing the list of 'players of the year' chosen annually since 1948 by the Football Writers' association. Table 8.3 gives the number of awards to players born in each country. It also measures the contribution of each country by taking the number of these awards per million of the country's population. We see that, not only do players move to England, but the smaller countries also produce substantially more of the top players than we would expect from their populations.

The latest development has been the rapid increase in the number of outstanding players from abroad, particularly from Continental Europe. For the years 1995 to 1999 the Football Writers' choices were Klinsmann, Cantona, Zola, Bergkamp and Ginola.

Chapter 9

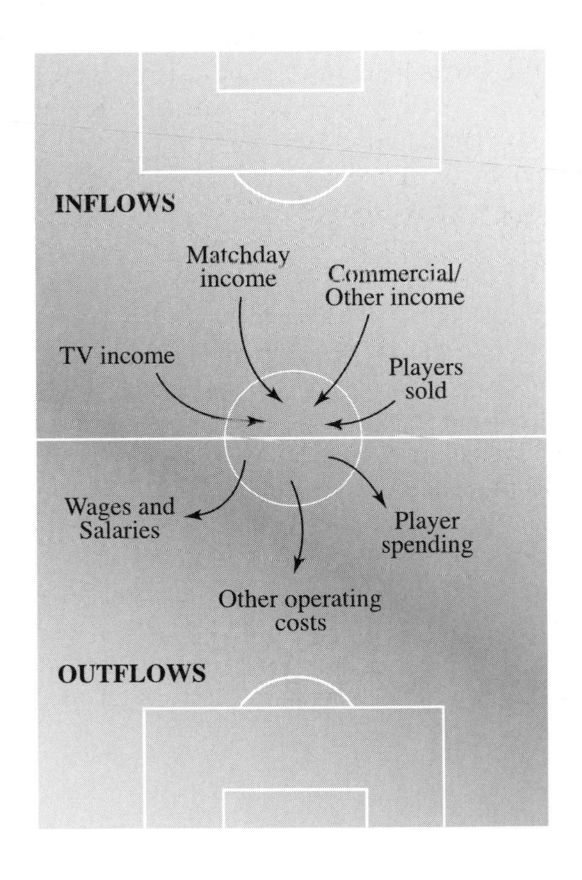
INFLOWS
Matchday income
Commercial/ Other income
TV income
Players sold
Wages and Salaries
Player spending
Other operating costs
OUTFLOWS

9

Economics

When modern football started in England in the middle of the nineteenth century the economics were very simple. The players usually had free access to a field, and the goal posts and playing kit could be bought by the players themselves.

The next stage arrived when it was found that football's popularity had grown to the point where spectators were willing to pay to watch it. The income so provided allowed clubs to attract players by giving them payments. For some years there was resistance to professionalism, but it was finally legalised in 1885.

It was not long before the clubs themselves expected a payment when a player moved to another club, leading to the development of the transfer system. This pattern persisted for many years and the economics remained quite straightforward.

Basically clubs with a large catchment area of potential spectators could achieve a good income from gate money. This was used to pay the players and support general expenses such as ground maintenance. Any remainder was available to buy players from other clubs. Transfer fees could provide a source of income for smaller clubs but generally the higher transfer fees were paid in transfers between larger clubs.

Until 1961 the full force of economic competition for players did not operate, there being a maximum wage which could be paid in each Division of the League. By present standards this maximum was incredibly low. Before the second world war it was typically three times average earnings. By the time the maximum wage was scrapped it had fallen to one and a half times. Today the top players have incomes a hundred times greater than the earnings of those who pay at the gate to watch them.

Over recent years the financing of professional football has changed dramatically, with new sources of income being exploited, particularly by the larger clubs. The first of these is sponsorship, the clubs being paid by a company to advertise its products, for example by carrying the company's name on the players' shirts. The second source of income is television. It was realised that the viewing public was eager to watch more football on television, and the introduction of satellite and cable television allowed this market to be tapped. The Premiership was able to negotiate a fee which originally was quite modest but has risen to tens of millions of pounds per club. Finally there is merchandising. There has been an unexpected enthusiasm of supporters, particularly the young, to buy replica football kits and other items carrying their club's name. The change is evident from a breakdown of the average Premiership club's turnover.

Match day receipts	37%
Television	29%
Commercial etc.	34%

Rather surprisingly the smaller clubs also receive most of their income from sources other than gate receipts. A typical breakdown is

Match day receipts	48%
Television	13%
Commercial etc.	39%

Size and success

The success of a football club depends on a number of factors but most directly on the ability of its players. In professional football this is related to the club's income since the more able players cost more money in transfer fees and command high wages. The club's income, in turn, depends on several factors but the basic element is the level of spectator support available to the club. It is quite obvious that a small town cannot compete with cities such as Manchester and Liverpool which have catchment areas with over a million people. Although, as we have seen, the gate money is only part of the club's income, it is also an indicator of the potential for income from commercial sales and other sources.

One measure of the support available to clubs is the attendance at matches. Let us start our analysis by looking at the relationship between success and attendance. The success of a club will be measured by taking its rank in the league tables averaged over three years. Thus the top team in the Premiership is ranked 1, the top team in the First Division is ranked 21 and the bottom team in the Third Division is ranked 92. Figure 9.1 gives the plot of attendance against rank.

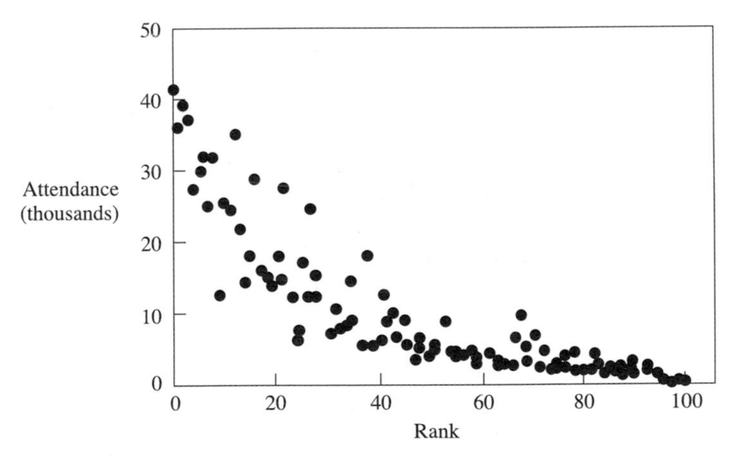

Figure 9.1. Graph of average attendance against the club's rank.

The correlation of attendance and rank is clear from the figure. However this, by itself, is not convincing evidence that high attendance produces a high rank since the correlation arises also from the fact that successful clubs attract greater support. These effects cannot be separated using the attendance/rank relation alone.

A more fundamental determining factor is the catchment area for potential support. This is, of course, difficult to define, but we can look at the broad trend by comparing rank with population. In the case of the large cities with wide surrounding areas of population, a mean of the populations of the city itself and of its broader conurbation area has been used. For each town only the highest ranked club has been included. London is obviously a complication because of its size and the large number of clubs, and is therefore excluded. Using this procedure a plot of rank against population is given in figure 9.2.

There is a wide spread of points in the graph, showing that small towns can be ambitious and that some large

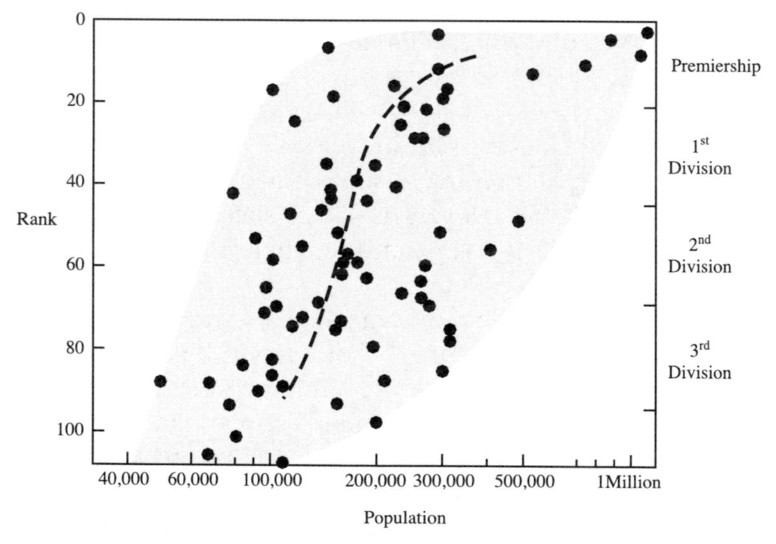

Figure 9.2. Graph of club's rank against size of town's population.

towns, such as Bristol, do not reach their potential. The graph does indicate the best a town can reasonably hope for, with a sufficiently large population being needed to achieve a place in each Division. As a rough guide the required populations are

	Minimum population (thousands)	Average population (thousands)
Premiership	100	300
1st Division	80	180
2nd Division	70	160
3rd Division	45	130

The minimum population is that required to reach each Division, and the average is the middle value for the Division. Of course the advent of a multi-millionaire benefactor can broaden a town's horizons.

Transfer fees

Nowadays the usual way that upper echelon clubs look to improve their teams is by paying transfer fees to acquire better players. The extent to which the club is able to do this depends on its income. The judgement as to how much of this income to spend on transfers is something of a balancing act. If buying better players leads to success and a higher income to balance the expenditure, that is fine. If not, the club can be in trouble.

The first thousand pound transfer fee was paid by Sunderland to Middlesborough for Alf Common in 1905. The British record fee has risen over the years to reach the £23.5 million paid by Manchester United for Juan Veron in 2001.

Figure 9.3 gives a graph of the British record transfer fee over almost a century. The early values are not resolved in the graph and it is useful therefore to move to a logarithmic scale. The resulting graph is shown in figure 9.4. The slight upward

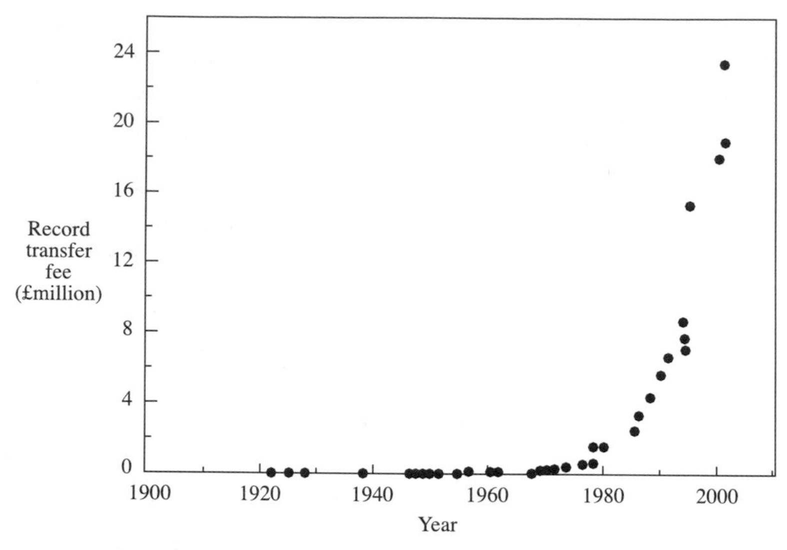

Figure 9.3. Graph of record transfer fee against time.

curvature of the graph shows that overall growth is somewhat faster than exponential. However, over the past 50 years the growth has been approximately exponential, fees doubling every 5 years. One wonders how long can this continue?

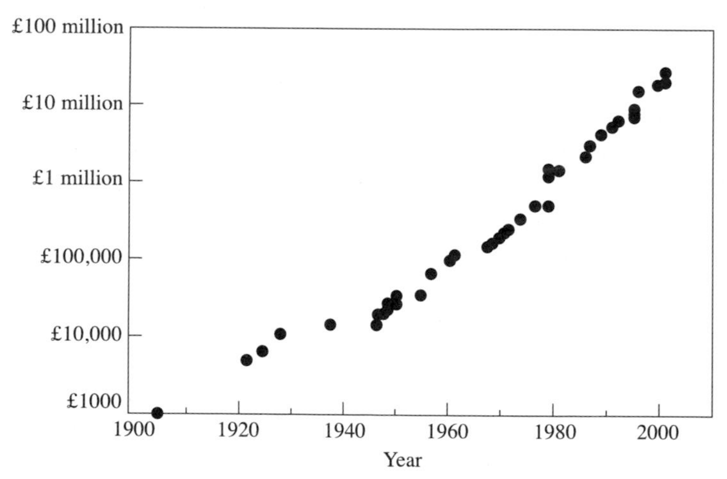

Figure 9.4. Graph of record transfer fee, plotted logarithmically, against time.

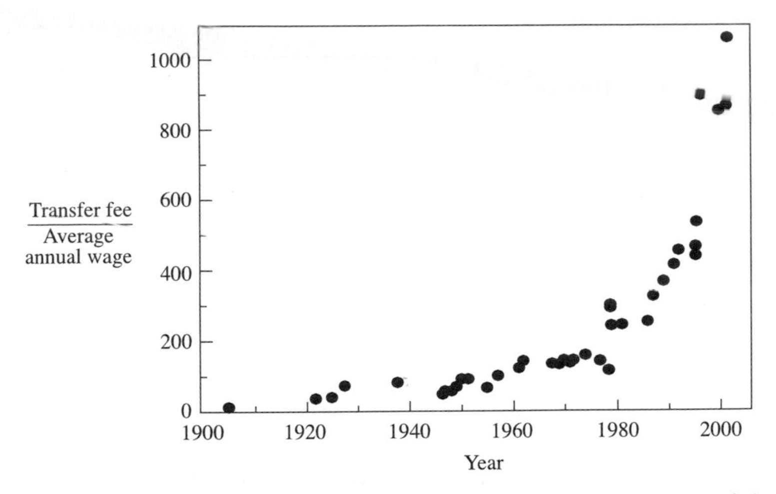

Figure 9.5. Graph of record transfer fee, in terms of the average wage of the general population.

Part of the growth in transfer fees results from the fall in value of the currency, inflation having reduced the value of the pound by a factor of 70 during this period. The general standard of living has also improved during this time as reflected in the growth in the real value of average earnings. A graph of the record transfer fee measured in terms of the average annual earnings of the time is given in figure 9.5.

The graph shows a remarkable growth. Alf Common was bought in 1905 for 13 years average earnings. It took more than a thousand years of average earnings to buy Juan Veron. Much of the growth has taken place in the past 20 years, during which the extra sources of income have become available to clubs.

Transfer fees make a big impact on the finances of some clubs, particularly the larger ones. Table 9.1 gives the average net amount of transfer fees per club in each Division as a percentage of the average turnover per club for a typical year. It is seen that the expenditure on transfer fees for Premiership clubs is quite substantial. For the First and Second Division clubs there is a small net income and for

Table 9.1

Division	Transfer payments as % of turnover
Premiership	−31%
First	+7%
Second	+7%
Third	+15%

the Third Division a somewhat more significant income, being 15% of turnover.

These figures cover a wide variation among the clubs. While some Premiership clubs have a low transfer expenditure, for others the cost can be more than the turnover of the whole Third Division.

Players' wages

The temptation for clubs in the lower Divisions is to buy players to achieve promotion. This is particularly true in the First Division where the rewards of the Premiership provide a great incentive. However, not only does the purchase of good players cost the transfer fees, it implies a continual drain on resources through the payment of wages. It is not uncommon for clubs to have a wage bill which exceeds the club's turnover. This clearly involves a gamble on the part of these clubs.

Interestingly Premiership clubs generally spend a smaller percentage of their turnover on wages than those in the lower divisions. Nevertheless the average Premiership expenditure on wages is more than half their turnover and many players now have million pound annual wages.

Chapter 10

LEGES MOTUS.

I.

Corpus omne perſeverare in ſtatu ſuo quieſcendi vel movendi uniformiter in directum, niſi quatenus illud a viribus impreſſis cogitur ſtatum ſuum mutare.

II.

Mutationem motus proportionalem eſſe vi motrici impreſſæ, et fieri ſecundum lineam rectam qua vis illa imprimitur.

III.

Actioni contrariam ſemper et equalem eſſe reactionem: ſive corporum duorum actiones in ſe mutuo ſemper eſſe æquales et in partes contrarias dirigi.

10

Mathematics

This chapter presents the mathematical calculations which underlie the models and examples given in the earlier chapters. The mechanics of the ball's behaviour are based on Newton's laws of motion and in particular the second law which states that the rate of change of momentum of a body is equal to the applied force. For us this usually takes the form force = mass × acceleration, but where rotation is involved it is more appropriate to describe the motion in terms of the change of angular momentum brought about by an applied torque. The models described all make the maximum use of simplifying assumptions to make the calculations as transparent as possible.

The account of the aerodynamics of the ball follows standard procedures for dealing with drag and the Magnus force. For anything but the simplest problems it is not possible to obtain algebraic solutions of the equations involved and the examples given in chapter 4 are the result of numerical calculations.

The equations for probabilities are given without derivation, which would be out of place here. However, their application is straightforward and the reader might find it interesting to substitute numbers for trial cases to check that they agree (or not) with their intuition.

The subjects dealt with are listed below, the first index number referring to the corresponding chapter.

1.1. Ideal bounce

During a bounce the ball initially undergoes an increasing deformation as the bottom surface is flattened against the ground. The resulting force, F, on the ball is given by the product of the excess air pressure, p, in the ball and the area of contact A, that is

$$F = pA. \tag{1}$$

For velocities of interest the deformation is sufficiently small that we can neglect the change in air pressure during the

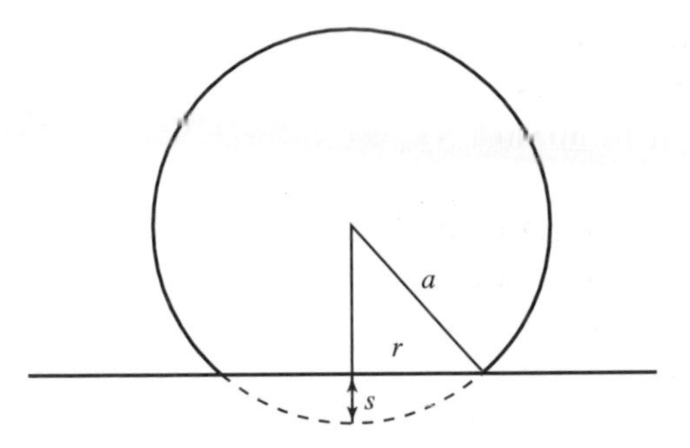

Figure 10.1. Geometry of the deformation.

bounce. In addition we shall initially neglect the frictional losses.

Figure 10.1 shows the geometry of the deformation, where a is the radius of the ball, s is the deformation depth and r is the radius of the circular surface of the ball in contact with the ground. From Pythagoras's theorem

$$a^2 = r^2 + (a - s)^2$$

so that

$$r^2 = 2as - s^2.$$

Usually s is sufficiently small that we can neglect the s^2 term and write the area

$$A = \pi r^2 = 2\pi as. \tag{2}$$

During the bounce the vertical velocity, v, of the centre of the ball is related to s by

$$v = -\frac{ds}{dt}. \tag{3}$$

The motion is described by Newton's second law and for an ideal bounce this takes the form

$$m\frac{dv}{dt} = F \tag{4}$$

where m is the mass of the ball. Combining equations (1) to

(4), we obtain the equation of motion

$$\frac{d^2 s}{dt^2} = -\frac{cp}{m} s \tag{5}$$

where c is the circumference of the ball, $2\pi a$. The solution of equation (5) is

$$s = \frac{v_0}{(cp/m)^{1/2}} \sin\left(\sqrt{\frac{cp}{m}}\, t\right) \tag{6}$$

where $t = 0$ is the time of the initial contact and v_0 is the magnitude of the vertical velocity of the ball at initial contact. At the time the ball leaves the ground, $s = 0$ again and this occurs when

$$\sqrt{\frac{cp}{m}}\, t = \pi$$

giving the duration of the bounce

$$t_b = \pi \sqrt{\frac{m}{cp}}. \tag{7}$$

We notice that, with our assumptions, the duration of the bounce does not depend on the initial velocity of the ball. Indeed it only depends on the mass, circumference and pressure of the ball, all of which are specified by the rules. Taking the average of the values allowed by the rules

$$m = 15 \text{ ounces} = 0.43\,\text{kg}$$

$$c = 27.5 \text{ inches} = 0.70\,\text{m}$$

$$p = 0.85 \text{ atmospheres} = 0.86 \times 10^5 \text{ Newtons m}^{-2},$$

equation (7) gives the bounce time $t_b = 8.4$ milliseconds, which is just under a hundredth of a second.

The maximum deformation depends on v_0 and occurs at $t = t_b/2$. From equations (6) and (7) its magnitude is

$$s_{max} = \frac{v_0 t_b}{\pi},$$

and substituting $t_b = 8.4 \times 10^{-3}$ seconds

$$s_{max} = 2.7 \times 10^{-3} v_0 \text{ metres} \qquad v_0 \text{ in m s}^{-1}.$$

Since $v_0(\text{m s}^{-1}) = 0.45 v_0(\text{mph})$ and $1 \text{ m} = 39.4$ inches

$$s_{max} = \frac{v_0}{21} \text{ inches} \qquad v_0 \text{ in mph}.$$

For example, a ball reaching the ground at 20 miles per hour would have a deformation of about an inch.

The maximum force on the ball occurs at maximum deformation. This occurs at $t = t_b/2$ and, from equations (3), (4), (6) and (7),

$$\begin{aligned} F_m &= \frac{\pi m v_0}{t_b} \\ &= 160 v_0 \text{ Newtons} \qquad v_0 \text{ in m s}^{-1} \\ &= 72 v_0 \text{ Newtons} \qquad v_0 \text{ in mph.} \end{aligned}$$

Since

$$1 \text{ Newton} = 0.102 \text{ kg wt} = 0.225 \text{ lbs wt} = 1.00 \times 10^{-4} \text{ tons}$$

the maximum force can be written

$$F_m = \frac{v_0}{140} \text{ tons} \qquad v_0 \text{ in mph.} \qquad (8)$$

1.2. Inelastic bounce

The assumption of a perfect bounce was quite adequate to obtain an approximate estimate of the bounce time and the deformation of the ball, but obviously cannot be used to describe the change of energy and spin brought about by the bounce.

When a ball bounces from a hard surface some of its kinetic energy is lost in inelastic deformation of the ball. In the case of a football on grass there is a further loss due to bending of the blades of grass, the loss depending on the length of the grass. Quantitatively this loss is measured by the coefficient of restitution, e, which is determined by the change of speed for a ball impacting a surface at a right angle. The definition is

$$e = \frac{\text{speed after impact}}{\text{speed before impact}}.$$

Because of the dependence on the playing surface this coefficient is quite variable, but on a good pitch it is typically around 0.5. The effect of the change of speed can be seen from the height of successive bounces. The height, h, of a bounce is found by equating the kinetic energy $\frac{1}{2}mv^2$ when leaving the ground to the potential energy mhg when the ball reaches the top of its bounce, g being the gravitational acceleration. Thus

$$h = \frac{v^2}{2g}.$$

If the ball now falls back to the ground it will again have a speed v on reaching the ground, but on leaving the ground after its second bounce it will have a velocity ev, and will now only bounce to a height h_2 given by

$$h_2 = \frac{(ev)^2}{2g} = e^2 h.$$

We see therefore that for $e = 0.5$ successive bounces are reduced to $\frac{1}{4}$ the height of the previous bounce. Players generally find this to be satisfactory. When plastic pitches were introduced into professional football for a while, they sometimes produced too high a bounce, making it more difficult to play a controlled game.

1.3. Angular momentum

Bounces usually involve spin and to investigate the role of spin it is necessary to introduce the concept of angular momentum. We shall take a brief diversion to look at this and to illustrate the basic elements involved in rotational motion.

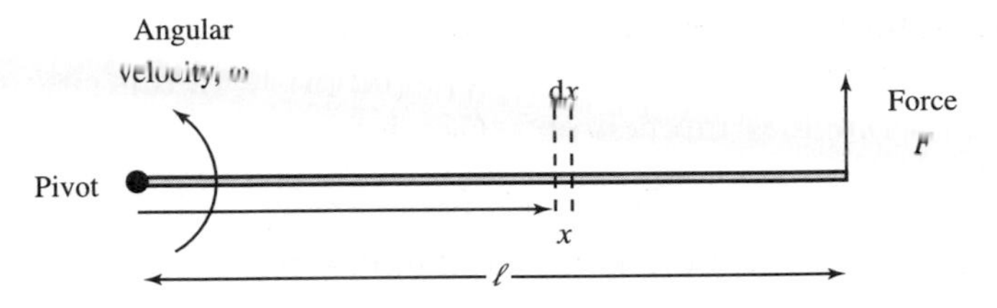

Figure 10.2. Pivoted rod.

For rotation about a fixed axis it is convenient to express Newton's second law in a form which gives the change of rotation in terms of the applied force. In this form the equations say that the rate of change of the angular momentum is equal to the applied torque. To understand these concepts, consider the simple example of a thin rod pivoted about one end, with a perpendicular force applied to the other, as illustrated in figure 10.2. For simplicity we shall assume there is no gravitational force. Let the rod have a varying mass distribution along its length, giving it a density ρ per unit length.

The energy of the rod is

$$E = \int_0^\ell \tfrac{1}{2}\rho v^2 \,\mathrm{d}x,$$

and since the velocity $v = \omega x$, where ω is the angular velocity,

$$E = \tfrac{1}{2} I\omega^2 \tag{9}$$

where

$$I = \int_0^\ell \rho x^2 \,\mathrm{d}x.$$

The quantity I is called the moment of inertia.

The rate of change of energy is given by the rate of work done by the force F. This is equal to the force times the velocity at its point of application, that is

$$\frac{\mathrm{d}E}{\mathrm{d}t} = F\mathrm{v} = F\ell\omega = \tau\omega. \tag{10}$$

The quantity τ, called the torque, is the product of the perpendicular force and its distance from the pivot, in this case $F\ell$.

The angular momentum, J, is defined as

$$J = I\omega$$

and from equation (9) its rate of change is given by

$$I\frac{\mathrm{d}\omega}{\mathrm{d}t} = \frac{1}{\omega}\frac{\mathrm{d}E}{\mathrm{d}t}.$$

Using equation (10) we now obtain the required equation of motion

$$I\frac{\mathrm{d}\omega}{\mathrm{d}t} = \tau. \tag{11}$$

This result applies more generally to all rigid bodies, each body with its specific mass distribution having a moment of inertia, I, for rotation about a given axis. Equation (11) then gives the change of rotation which results from a torque τ.

1.4. Bounce at an angle

Having examined the vertical bounce of a ball without spin we now turn to the general case in which a spinning ball strikes the ground at an angle. If the ball bounces on a rough surface its spin will change during the bounce, and even a ball without spin will acquire a spin during the bounce.

First let us define the quantities involved in the bounce. Figure 10.3 indicates the velocity components and spin before and after the bounce.

In the diagram the ball bounces from left to right and a clockwise spin is taken to be positive. The angular velocities before and after the bounce are ω_0 and ω_1. The corresponding horizontal velocities are u_0 and u_1, and the vertical velocities are v_0 and v_1. It should be noted that the initial vertical velocity v_0 is here taken to be positive.

The analysis of the bounce is different for the cases where the ball slides throughout the bounce, and where the ball is

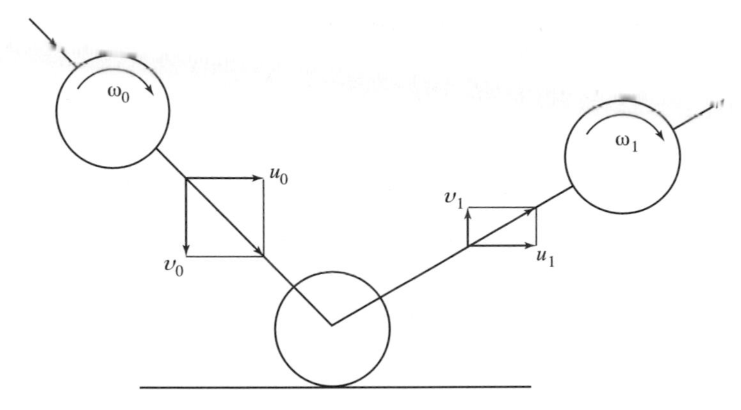

Figure 10.3. Showing the conditions before and after the bounce.

rolling on leaving the bounce. We shall consider these cases in turn. However, one aspect of the bounce is common to both – the vertical velocities are related by the coefficient of restitution, and

$$v_1 = ev_0. \tag{12}$$

Consequently the change in vertical velocity, Δv, from v_0 downwards to v_1 upwards is given by

$$\Delta v = v_1 - (-v_0) = v_0 + v_1 = (1 + e)v_0. \tag{13}$$

1.5. Bounce with ball sliding

If the ball slides throughout the bounce there is a horizontal friction force, F_h, acting on the bottom of the ball as illustrated in figure 10.4. This force slows the ball and also imposes a torque $F_h a$ about the centre of gravity where a is the radius of the ball. The friction force is given by

$$F_h = \mu F_v \tag{14}$$

where μ is the coefficient of sliding friction and F_v is the vertical force between the ball and the ground.

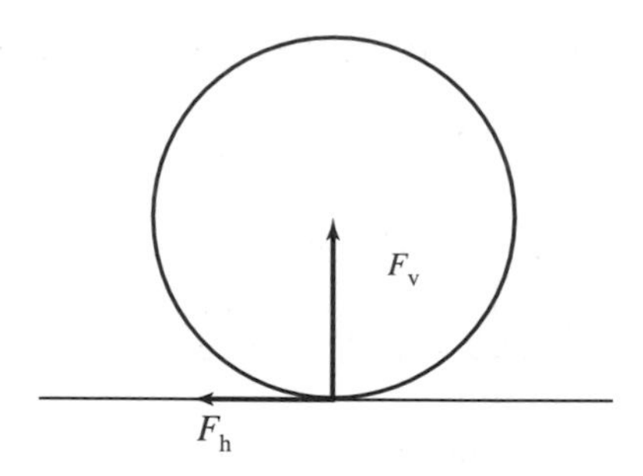

Figure 10.4. Friction force F_h resulting from vertical force F_v.

Newton's second law gives the equations for the horizontal and vertical velocities during the bounce

$$m\frac{du}{dt} = -F_h \quad \text{and} \quad m\frac{dv}{dt} = F_v \tag{15}$$

so that

$$\frac{du}{dv} = -\frac{F_h}{F_v} \tag{16}$$

and the change in the horizontal velocity, $\Delta u = u_1 - u_0$, during the bounce is given by integrating equation (16) through the bounce using equation (14). This gives

$$\Delta u = -\mu \Delta v,$$

and using equation (13)

$$\Delta u = -\mu(1 + e)v_0. \tag{17}$$

The change in rotation due to the force, F_h, is given by the equation of motion (11)

$$I\frac{d\omega}{dt} = F_h a \tag{18}$$

where I is the moment of inertia of the ball and $F_h a$ is the torque. Equations (15) and (18) give

$$\frac{d\omega}{dt} = -\frac{ma}{I}\frac{du}{dt}$$

and integrating this equation, the change in ω is

$$\Delta\omega = -\frac{ma}{I}\,\Delta u. \tag{19}$$

Substitution of equation (17) into equation (19) gives

$$\Delta\omega = \mu(1+e)\,\frac{ma}{I}\,v_0. \tag{20}$$

The moment of inertia of a hollow sphere about an axis through its centre is

$$I = \tfrac{2}{3}ma^2$$

and substituting this relation into equation (20) gives the change of rotation frequency during the bounce

$$\Delta\omega = \frac{3}{2}\,\mu(1+e)\,\frac{v_0}{a}. \tag{21}$$

Summarising these results, equations (12), (17) and (21) give the velocities and rotation resulting from a sliding bounce

$$v_1 = ev_0, \qquad u_1 = u_0 - \mu(1+e)v_0 \tag{22}$$

$$\omega_1 = \omega_0 + \frac{3}{2}\,\mu(1+e)\,\frac{v_0}{a}. \tag{23}$$

1.6. Bounce with ball rolling

When the ball touches the ground and slides, the friction force, F_h, on the ball slows the lower surface. For rougher surfaces and for higher angles of approach the force brings the lower surface to a halt and the ball then rolls through the bounce as illustrated in figure 10.5.

In this case equation (14), describing the sliding friction force, is no longer applicable. It is replaced by the condition that the ball finishes the bounce rolling, that is

$$u_1 = \omega_1 a. \tag{24}$$

The other relationship between u_1 and ω_1 comes from

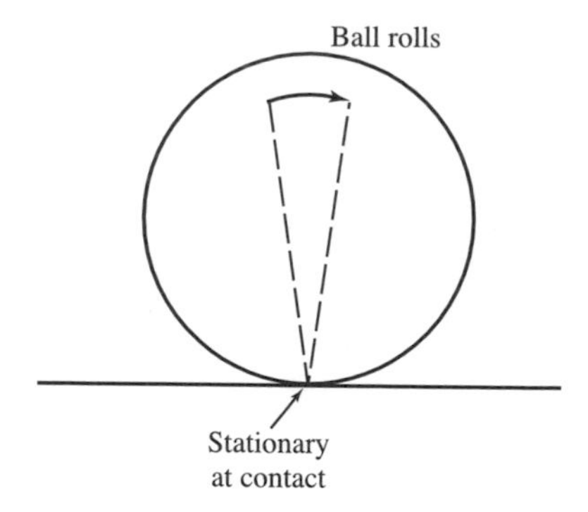

Figure 10.5. Ball rolling during bounce.

equation (19), and since $I = \frac{2}{3}ma^2$ this gives

$$\omega_1 - \omega_0 = -\frac{3}{2}\frac{u_1 - u_0}{a}. \tag{25}$$

Equation (12), giving the change in vertical velocity, still holds and equations (24) and (25) together with equation (12) give the conditions resulting from the rolling bounce.

$$v_1 = ev_0 \tag{26}$$

$$u_1 = \frac{3}{5}u_0 + \frac{2}{5}\omega_0 a \tag{27}$$

$$\omega_1 = \frac{2}{5}\omega_0 + \frac{3}{5}\frac{u_0}{a}. \tag{28}$$

1.7. Condition for rolling

The rolling relation given by equation (24) can be written $u_1/\omega_1 a = 1$. Provided the ratio $u_1/\omega_1 a$ predicted by the 'sliding' equations (22) and (23) is greater than 1 the bounce is in the sliding regime. If the equations predict $u_1/\omega_1 a < 1$ they are no longer valid and the bounce is in the rolling regime. Using equations (22) and (23) this gives the condition for rolling to take place

$$\mu(1 + e)v_0 > \tfrac{2}{5}(u_0 - \omega_0 a). \tag{29}$$

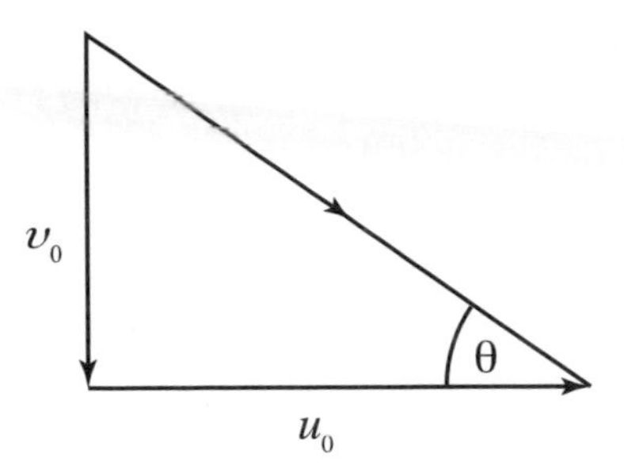

Figure 10.6. Tan $\theta = v_0/u_0$.

If the ball is not spinning before the bounce the condition for rolling becomes simply a requirement that the angle of approach to the bounce, θ, be sufficiently large. From figure 10.6, $\tan\theta = v_0/u_0$ and so, from inequality (29), the condition for rolling becomes

$$\tan\theta > \frac{2}{5\mu(1+e)}.$$

For example if $\mu = e = 0.7$, rolling occurs for $\theta > 19°$.

1.8. Angle of rebound

The angle of rebound can be calculated from the vertical and horizontal components of the velocity which we have already determined. The geometry is shown in figure 10.7.

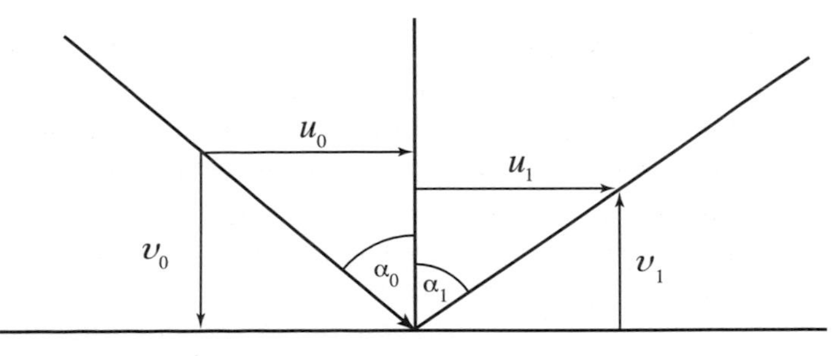

Figure 10.7. Geometry of bounce.

The angle of rebound to the vertical, α_1, is given by

$$\tan \alpha_1 = \frac{u_1}{v_1}$$

and using equations for the case where the ball slips

$$\tan \alpha_1 = \frac{u_0 - \mu(1+e)v_0}{ev_0}.$$

Since

$$\frac{u_0}{v_0} = \tan \alpha_0$$

we have the relation of the angle of rebound to the angle of incidence, α_0,

$$\tan \alpha_1 = \frac{1}{e} \tan \alpha_0 - \mu\left(1 + \frac{1}{e}\right).$$

Similarly for the case of a bounce where the ball leaves the ground rolling, equations (26) and (27) give

$$\tan \alpha_1 = \frac{3}{5e} \tan \alpha_0 + \frac{2}{5e} \frac{\omega_0 a}{v_0}. \qquad (30)$$

1.9. Rebound from the crossbar

The geometry of the bounce from the crossbar is shown in figure 10.8. ϕ_0 and ϕ_1 are the angles of the ball's velocity to the horizontal, before and after impact.

There are two parts to the calculation of the bounce. Firstly we use the results of the previous section to determine the relationship of the angles of incidence and rebound. In this case the surface from which the bounce takes place is replaced by the tangent AB through the point of contact. The second part of the calculation relates the angle of this tangent to the height of the ball at the bounce in relation to the position of the bar. The ball will actually move on the bar during the bounce, but to keep the calculation simple we shall take the

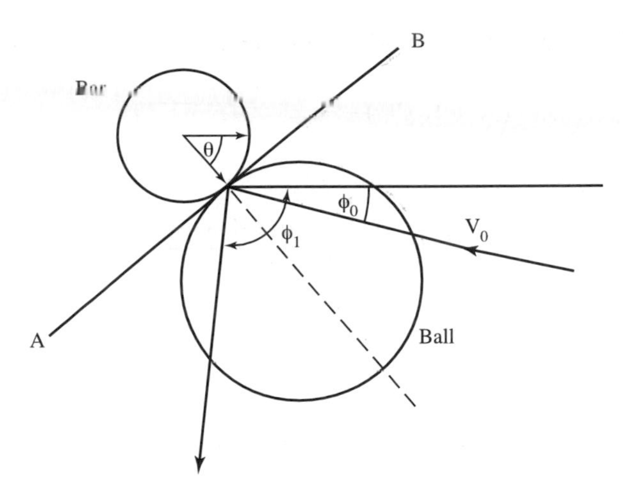

Figure 10.8. Geometry of bounce from the crossbar.

contact position on the bar to be that in the middle of the bounce.

From figure 10.8 the angle of incidence is

$$\alpha_0 = \theta - \phi_0$$

and the angle of the rebound is

$$\alpha_1 = \phi_1 - \theta.$$

Taking the ball to be rolling from the bounce, α_1 and α_0 are related by equation (30). Assuming, for simplicity, that the ball is not spinning before the bounce, this gives an equation for ϕ_1

$$\tan(\phi_1 - \theta) = \frac{3}{5e}\tan(\theta - \phi_0). \tag{31}$$

It now remains to relate θ to the height at which the ball bounces on the bar. The geometry is shown in figure 10.9.

If the radius of the ball is a and the radius of the bar is b, the difference in height, h, between the centre of the bar and centre of the ball is

$$h = (a + b)\sin\theta. \tag{32}$$

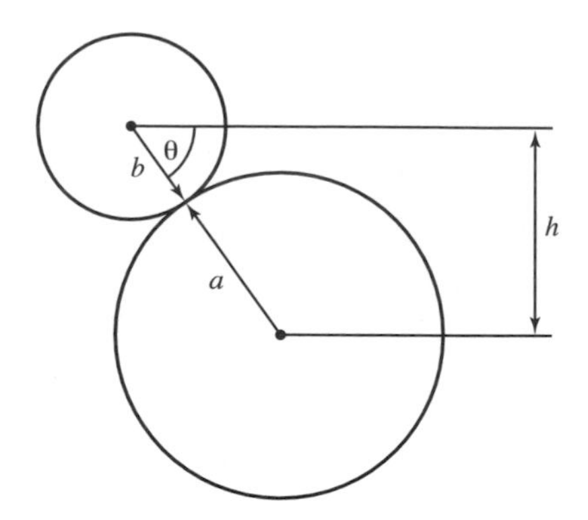

Figure 10.9. Relating h to a, b and θ.

Thus for a given h, equation (32) determines θ, and using this value in equation (31) gives the angle of a rebound ϕ_1, given the angle of incidence, ϕ_0.

To calculate the rotation of the ball after the rebound we use equation (28). To do this we need an equation for u_0. From figure 10.8 the angle between the incoming velocity, V_0, and the normal to the line AB is $\theta - \phi_0$. The required tangential velocity u_0 is therefore given by

$$u_0 = V_0 \sin(\theta - \phi_0)$$

and, from equation (28), the rotation frequency after the bounce, with $\omega_0 = 0$, is

$$\omega_1 = \frac{3}{5}\frac{V_0}{a}\sin(\theta - \phi_0)$$

where θ is given by equation (32).

2.1. The kick

In a hard kick the leg is swung like a double pendulum, pivoted at the hip and jointed at the knee. The leg is first accelerated and then decelerated to rest. The ball is struck close to the time of maximum velocity, and at this time the leg is almost straight. Essentially the ball bounces off the moving foot. Since this bounce takes some time the ball remains in contact with the foot for a finite distance. For a kick in which the foot is moving at 50 miles per hour with a bounce time of one hundredth of a second, contact is maintained for about 9 inches, roughly the diameter of the ball.

The mechanics of the kick are rather complex but we can simplify the analysis by assuming that during contact with the ball the leg just pivots about the hip. When the foot has reached its maximum velocity the process is then that of transferring momentum from the leg to the ball. If the leg, including the foot, has a moment of inertia I about the hip, its angular momentum at the start of impact is $I\Omega_0$, where Ω_0 is the initial angular velocity of the leg. At the end of the impact the angular velocity is reduced to Ω_1 and the angular momentum is $I\Omega_1$. The lost angular momentum is transferred to the ball whose angular momentum about the hip is $m\ell v_{\mathrm{b}}$ where m is the mass of the ball, ℓ the length of the leg and v_{b} is the velocity given to the ball. Thus

$$I(\Omega_0 - \Omega_1) - m\ell v_{\mathrm{b}}$$

and writing the initial velocity of the foot as $v_0 = \Omega_0 \ell$, and the velocity after impact as $v_1 = \Omega_1 \ell$

$$I(v_0 - v_1) = m\ell^2 v_{\mathrm{b}}. \tag{33}$$

If we describe the bounce of the ball from the foot in terms of a coefficient of restitution e,

$$(v_{\mathrm{b}} - v_1) = e v_0. \tag{34}$$

Then, using equation (34) to eliminate v_1 in equation (33), we obtain the velocity of the ball in terms of the initial velocity of the foot

$$v_{\mathrm{b}} = v_0 \frac{1 + e}{1 + (m\ell^2/I)}. \tag{35}$$

Because the mass of the leg is much greater than that of the ball, I is several times $m\ell^2$ and consequently $m\ell^2/I$ is less than e. This means that the ball leaves the foot with a higher velocity than the velocity of the foot.

Using equations (33) and (35) the fractional change in the velocity of the foot is

$$\frac{v_1 - v_0}{v_0} = -\frac{1+e}{1+(I/m\ell^2)}$$

and since $I/m\ell^2 \gg 1$, this shows that the foot is only slightly slowed by the impact with the ball.

3.1. The throw

For a throw-in a continuous force is applied to the ball as it is moved forward together with the hand and arms. The momentum which can be given to the ball is limited by the distance the arms can be moved before the ball is released. If a constant force, F, were applied for a time t, the acceleration F/m would produce a velocity

$$v = \frac{Ft}{m} \tag{36}$$

and, since the distance covered is $d = \int v\,\mathrm{d}t$,

$$d = \frac{Ft^2}{2m}. \tag{37}$$

Equations (36) and (37) give the velocity achieved over the distance d

$$v = \sqrt{\frac{2Fd}{m}}. \tag{38}$$

However, as the arms move forward and the ball speeds up it becomes difficult to maintain the force and the acceleration. The force starts at a high value and probably falls close to zero if the arms are extended well forward. Thus, for long throws the force appearing in equation (38) must be replaced by an average value. For short throws contact with the ball is only maintained for a short distance. For a given applied force this distance falls off as the square of the required velocity.

When the ball is hurled by the goalkeeper the same equations apply but the distance over which the force can be maintained is longer.

3.2. The catch

Since a catch is the inverse of a throw it is described by the same equations. However, in this case it is the initial velocity, v, which is known, and for a given take-back distance, d, of the hands, equation (38) gives the average force on the hands

$$F = \frac{\frac{1}{2}mv^2}{d}.$$

This equation brings out the fact that the decelerating force applied by the hands is that necessary to remove the kinetic energy, $\frac{1}{2}mv^2$, of the ball in the distance d.

4.1. Flight of the ball

The flight of the ball is determined by Newton's second law of motion

$$\text{force} = \text{mass} \times \text{acceleration}.$$

In the general case there are three forces acting on the ball, the force of gravity and two forces arising from interaction with the air. The simplest force from the air is drag, which acts in the opposite direction to the ball's velocity. The other, more subtle, force is the Magnus force which, in the presence of spin, acts at right angles both to the velocity and to the axis of spin. With spin about a horizontal axis the Magnus force can provide lift; with spin about a vertical axis the flight of the ball is made to bend.

When the effect of the air is negligible the equations of motion are easily solved. Since there is no horizontal force the equation for the horizontal velocity, u, is

$$m\frac{\mathrm{d}u}{\mathrm{d}t} = 0$$

and so the horizontal velocity is constant, and u is equal to the initial horizontal velocity u_0. The horizontal displacement, x, is therefore

$$x = u_0 t. \tag{39}$$

The equation for the vertical velocity, v, is

$$m\frac{\mathrm{d}v}{\mathrm{d}t} = -mg$$

where g is the acceleration due to gravity. This equation has the solution

$$v = v_0 - gt$$

where v_0 is the initial vertical velocity. Since $v = \mathrm{d}y/\mathrm{d}t$ the vertical displacement is obtained by integrating

$$\frac{\mathrm{d}y}{\mathrm{d}t} = v_0 - gt$$

to obtain

$$y = v_0 t - \tfrac{1}{2}gt^2. \tag{40}$$

Using equation (39) to eliminate t in equation (40) gives the equation for the trajectory

$$y = \frac{v_0}{u_0}x - \frac{1}{2}\frac{g}{u_0^2}x^2. \tag{41}$$

and this is the equation of a parabola.

The range of the flight is obtained by putting $y = 0$ in equation (41). Obviously $y = 0$ for $x = 0$, but the other solution for x gives the range

$$R = \frac{2v_0 u_0}{g}. \tag{42}$$

The time of flight is given by the time, $t = T$, at which the displacement y returns to zero. From equation (40) this is given by

$$T = \frac{2v_0}{g}.$$

If the initial angle between the trajectory and the ground is θ_0, then

$$v_0 = V_0 \sin\theta_0 \qquad \text{and} \qquad u_0 = V_0 \cos\theta_0 \tag{43}$$

where the initial total velocity, V_0, is given by

$$V_0^2 = v_0^2 + u_0^2.$$

In terms of V_0 and θ_0 the range given by equation (42) becomes

$$R = \frac{2V_0^2 \sin\theta_0 \cos\theta_0}{g}$$

and, using the identity $2\sin\theta_0\cos\theta_0 = \sin 2\theta_0$,

$$R = \frac{V_0^2 \sin 2\theta_0}{g}.$$

Since $\sin 2\theta_0$ has its maximum value at $\theta_0 = 45°$, this angle gives the maximum range for a given V_0,

$$R_{\text{max}} = \frac{V_0^2}{g}.$$

4.2. Flight with drag

The drag force on a body moving in air is conventionally written

$$F_\text{d} = \tfrac{1}{2} C_\text{D} \rho A V^2 \tag{44}$$

where the drag coefficient C_D depends on the velocity, ρ is the density of the air, V is the velocity of the body, and A is its cross-sectional area, in our case πa^2.

Although equation (44) is simple, the solution of the associated equations of motion is rather involved. This is partly because of the velocity dependence of C_D but is also due to the fact that the drag force couples the equations for the horizontal and vertical components of the velocity. Newton's equations now become

$$m\frac{du}{dt} = -F_d \cos\theta \tag{45}$$

and

$$m\frac{dv}{dt} = -F_d \sin\theta - mg \tag{46}$$

where θ is the angle between the trajectory and the ground at time t, given by

$$\tan\theta = \frac{v}{u}. \tag{47}$$

Even for constant C_D, equations (44) to (47) do not have an algebraic solution, but they are easily solved numerically for any particular case using a computer.

If C_D is taken to be a constant during the flight then, using $v = V\sin\theta$ and $u = V\cos\theta$, equations (45) and (46) can be conveniently written.

$$\frac{du}{dt} = -\alpha uV \tag{48}$$

$$\frac{dv}{dt} = -\alpha vV - g \tag{49}$$

where

$$V^2 = v^2 + u^2 \tag{50}$$

and

$$\alpha = \tfrac{1}{2} C_D \rho A/m.$$

In the calculations for the cases presented in chapter 4, equations (48) to (50) were solved with C_D taken to be 0.2. The density of air is 1.2 kg m^{-3}, the mass of the ball is

0.43 kg, and its cross-sectional area is 0.039 m^2, giving the value $\alpha = 0.011\ m^{-1}$.

Having solved for u and v it is straightforward to obtain x and y by integrating $dx/dt = u$ and $dy/dt = v$.

4.3. Effect of a wind

The drag on the ball is determined by its velocity with respect to the air. Thus for a wind having a velocity w along the direction of the ball's flight the equations of motion (48) and (49) take the form

$$\frac{du}{dt} = -\alpha(u - w)V \tag{51}$$

$$\frac{dv}{dt} = -\alpha v V - g \tag{52}$$

with V now given by

$$V^2 = (u - w)^2 + v^2. \tag{53}$$

A positive value of w corresponds to a trailing wind, and a negative value corresponds to a headwind.

Again, the equations can be solved directly using a computer. It is interesting to note, however, that if we make the transformation $u - w \longrightarrow u'$ with $v \longrightarrow v'$, equations (51) to (53) take the form of equations (48) to (50) with u and v replaced by u' and v'. If the equations are solved for u' and v', and x' and y' are calculated from $dx'/dt = u'$ and $dy'/dt = v'$, the required solutions can then be obtained using the inverse transformations.

$$u = u' + w \qquad v = v'$$

$$x = x' + wt \qquad y = y'.$$

This does not mean that the values of the vertical velocity and position, v and y, are unchanged by the wind since the wind-modified value of V enters into the calculation of v'. As usual, the range and time-of-flight are determined by the condition that the ball has returned to the ground, that is $y = 0$.

4.4. Effect of a sidewind

If there is a sidewind with velocity w, the motion in the direction, z, of this wind is obtained from the equation for the velocity, v_z, in this direction

$$\frac{\mathrm{d}v_z}{\mathrm{d}t} = -\alpha(v_z - w)V \tag{54}$$

with

$$V^2 = u^2 + v^2 + (v_z - w)^2.$$

Again this equation can be solved numerically together with the equations for u and v. However a simple procedure gives a formula for the sideways deflection of the ball's trajectory which is sufficiently accurate for most circumstances.

The equation for the forward motion is

$$\frac{\mathrm{d}u}{\mathrm{d}t} = -\alpha uV \tag{55}$$

and dividing equation (54) by equation (55) gives

$$\frac{\mathrm{d}v_z}{\mathrm{d}u} = \frac{v_z - w}{u}. \tag{56}$$

Integration of equation (56) gives the solution

$$v_z = w\left(1 - \frac{u}{u_0}\right) \tag{57}$$

where u_0 is the initial value of u and $v_z = 0$ initially.

The deflection z is obtained by solving

$$\frac{\mathrm{d}z}{\mathrm{d}t} = v_z.$$

Thus, using equation (57) for v_z

$$z = w\left(t - \frac{\int_0^t u\,\mathrm{d}t}{u_0}\right).$$

The deflection, *d*, over the full trajectory is therefore

$$d = w\left(T - \frac{R}{u_0}\right)$$

where *T* is the time of flight and *R* is the range. Since *T* and *R* are little affected by the sidewind, a good approximation for *d* is obtained using their values with no wind. If there were no air drag, then $T = R/u_0$ and deflection would, of course, be zero.

4.5. The Magnus effect

When the ball is spinning the Magnus effect produces a force on the ball which is perpendicular to the spin and perpendicular to the ball's velocity, as illustrated in figure 10.10. Conventionally this force is written

$$F_\mathrm{L} = \tfrac{1}{2} C_\mathrm{L} \rho A V^2$$

by analogy with the drag force given in equation (44). This formula has its origin in aeronautics and the subscript L

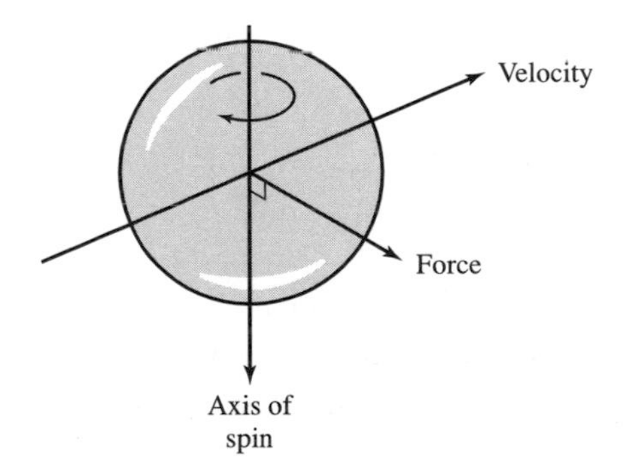

Figure 10.10. Illustrating the relation of the Magnus force to the ball's velocity and spin.

stands for the lift which would occur, for example, on a wing. For our purpose this expression is somewhat misleading because C_L depends on both the spin and the velocity.

For a spinning ball C_L is proportional to $\omega a/V$ provided $\omega a/V$ is not too large and it is, therefore, convenient to write

$$C_L = \frac{\omega a}{V} C_s$$

where ω is the angular frequency of the spin and a is the radius of the ball. Then

$$F_L = \tfrac{1}{2} C_s \rho A a \omega V. \tag{58}$$

Substituting for the air density, $\rho = 1.2\,\mathrm{kg\,m^{-3}}$, the radius $a = 0.11\,\mathrm{m}$ and the cross-sectional area $A = 0.039\,\mathrm{m^2}$, equation (58) becomes

$$F_L = 2.6 \times 10^{-3} C_s \omega V \text{ Newtons} \qquad V \text{ in m s}^{-1}. \tag{59}$$

This sideways force produces a curved trajectory and the force is balanced by the centrifugal force mV^2/R, where R is the radius of curvature of the trajectory. Using equation (59) with a mass of 0.43 kg, the resulting radius of curvature is

$$R = 165 \frac{V}{C_s \omega} \text{ metres} \qquad V \text{ in m s}^{-1}. \tag{60}$$

If we measure the rotation by the number of revolutions per second, f, then since $f = \omega/2\pi$, equation (60) becomes

$$R = 26 \frac{V}{C_s f} \text{ metres} \qquad V \text{ in m s}^{-1}. \tag{61}$$

It is more natural to think in terms of sideways displacement of the ball as illustrated in figure 10.11. If we approximate by taking the trajectory to have a constant curvature then using Pythagoras's equation

$$L^2 + (R - D)^2 = R^2$$

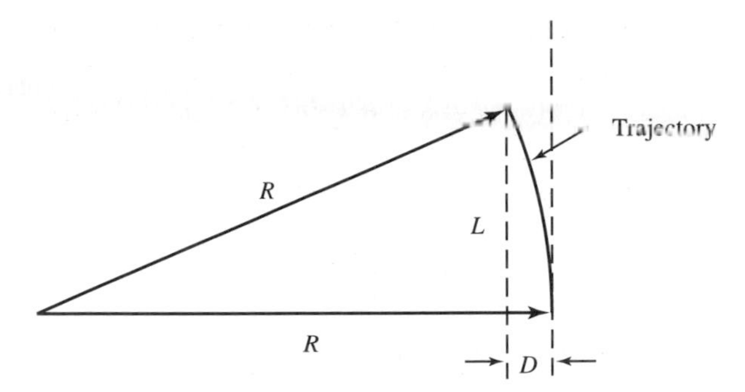

Figure 10.11. Deviation, D, arising from the ball's curved trajectory.

and, taking $D \ll R$ so that D^2 is negligible

$$D = \frac{L^2}{2R}.$$

Using equation (61) this becomes

$$D = \frac{C_s L^2 f}{52V} \text{ metres} \qquad V \text{ in m s}^{-1}. \qquad (62)$$

The time of flight is L/V and so the number of revolutions of the ball during its flight is $n = Lf/V$. Substitution of this relation into equation (62) gives

$$\frac{D}{L} = C_s \frac{n}{52}.$$

We have no direct measurement of C_s for footballs but experiments with other spheres have given values in the range $\frac{1}{4}$ to 1 depending on the nature of the surface. Taking $C_s = \frac{1}{2}$ we obtain the approximate relation

$$\frac{D}{L} = \frac{n}{100}.$$

For example, a deviation of 1 m over a length of 30 m would require the ball to undergo about 3 revolutions.

The ratio of f/V appearing in equation (62) is related to the ratio of the rotational energy to the kinetic energy. This ratio is

$$\frac{E_R}{E_K} = \frac{\frac{1}{2} I\omega^2}{\frac{1}{2} m V^2}$$

and since $I = \frac{2}{3} m a^2$

$$\frac{E_R}{E_K} = 0.32 \left(\frac{f}{V} \right)^2 \qquad V \text{ in m s}^{-1}.$$

For the example, a ball travelling at 30 mph ($13.4\,\mathrm{m\,s^{-1}}$) with a spin of 3 revolutions per second has a rotational energy of 1.6% of its kinetic energy.

4.6. Producing targeted flight with spin

In a normal kick the ball is kicked along a line through the centre of the ball and this means that the ball is struck at a right angle to its surface. If the flight of the ball is to be bent, the angle of the kick to the surface must be turned away from a right angle in order to apply a torque to the ball and give it spin. A further requirement is that the ball must be struck at the correct place on the surface, which is no longer on the line through the centre of the ball in the direction of the flight. Using the aerodynamics of the flight and the mechanics of the kick we can determine the necessary prescription. The calculation has five parts:

(i) The geometry of the flight.
(ii) Relating the spin and sideways velocity produced by the kick.
(iii) Relating the forward velocity of the ball to the velocity of the foot.
(iv) Application of the constraint that the ball moves with the foot.
(v) Combining the above calculations to obtain the required prescription.

We shall look at these parts in turn. For simplicity we shall take the angles involved to be small to avoid the introduction of trigonometric functions. To avoid too much complication we shall not include the change in the position of the foot on the ball during the kick and will take the position of the foot to be represented by its average position during the contact.

(i) Geometry of the flight

To place a curved shot on target requires that it be kicked in the correct direction with the required spin. The geometry of the flight is shown in figure 10.12.

The ball leaves the foot at an angle ϕ to the direction of the target and the trajectory has an initial direction aimed at a distance D from the target which is a distance L away. Taking the angle ϕ to be small, the required kick calls for a

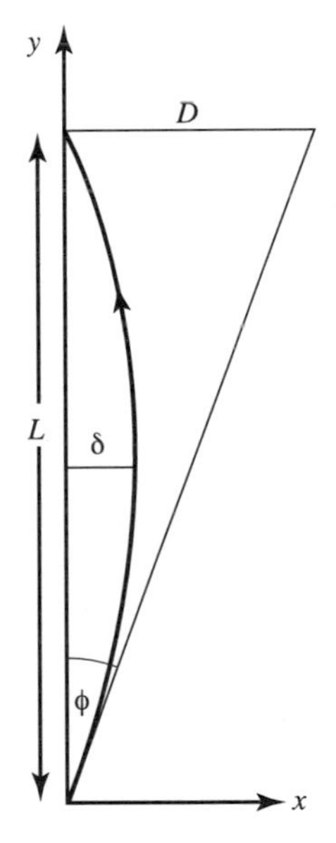

Figure 10.12. Geometry of the curved flight.

departure angle $\phi = D/L$. Using equation (58) for the force on the ball the equation of motion is

$$m\frac{d^2x}{dt^2} = -\frac{1}{2}C_s\rho A\omega aV.$$

Neglecting drag and using the approximation $y = Vt$, we obtain the equation for the ball's trajectory

$$x = \frac{1}{4}C_s\frac{\omega a}{V}\frac{L}{\ell}y\left(1 - \frac{y}{L}\right) \tag{63}$$

where $\ell = m/\rho A$ is the length over which the mass of air swept by the cross-sectional area A is equal to the mass of the ball, and $\omega a/V$ is the ratio of the equatorial spin velocity to the velocity of the ball.

The maximum deviation of the ball from the straight line to the target occurs at $y = L/2$ and is

$$\delta = \frac{1}{16}C_s\frac{\omega a}{V}\frac{L^2}{\ell}.$$

This equation gives the required spin, ω, for a given deviation. To produce this deviation the ball must be kicked towards a point at a distance D from the target where $D = 4\delta$, and the required spin is

$$\omega = \frac{4VD\ell}{C_s aL^2}. \tag{64}$$

The task of the kicker is now defined. To produce a deviation D with a ball kicked with a velocity V the ball must be kicked at the angle $\phi = D/L$, and be given a spin ω in accordance with equation (64).

The required angle, ϕ, can be related to the spin by substituting $D/L = \phi$ in equation (64) to obtain

$$\phi = \frac{1}{4}C_s\frac{\omega a}{V}\frac{L}{\ell}. \tag{65}$$

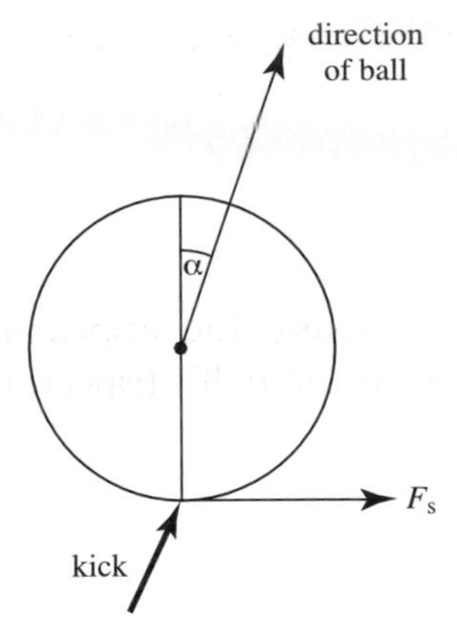

Figure 10.13. Geometry of the kick.

(ii) The kick with spin

To produce the spin required for a curled flight it is necessary to strike the ball 'off-centre' and at an angle as shown in figure 10.13.

The force of the kick has a sideways component $F_s(t)$ which gives the ball a velocity component $u(t)$ in the direction of F_s and, through the torque it applies, a spin $\omega(t)$. The equations for the transfer of linear and angular momentum are

$$m\frac{du}{dt} = F_s$$

and

$$I\frac{d\omega}{dt} = aF_s$$

where a is the radius of the ball and I is the moment of inertia about a diameter which, for a hollow sphere, is $\frac{2}{3}ma^2$. These equations combine to give

$$\frac{du}{d\omega} = \frac{2}{3}a$$

and so when the kick is completed the final values are related by

$$u = \tfrac{2}{3}\omega a. \tag{66}$$

This sideways velocity deflects the ball's direction away from the direction through the centre of the ball. Taking the deflection angle, α, to be small so that $\tan\alpha$ can be replaced by α, it can now be written

$$\alpha = \frac{u}{V} = \frac{2}{3}\frac{\omega a}{V}. \tag{67}$$

(iii) Velocity of the ball

The 'forward' motion is dealt with by introducing the coefficient of restitution. Taking the angle between the direction of the kick and the departure direction of the ball to be small the departure velocity of the ball is

$$V = (1 + e)v_{\mathrm{f}} \tag{68}$$

where v_{f} is the velocity of the foot.

Equations (67) and (68) combine to give the deflection angle for a given spin

$$\alpha = \frac{2}{3(1+e)}\frac{\omega a}{v_{\mathrm{f}}}. \tag{69}$$

(iv) The required spin

In the previous section we calculated the angle α for the direction of the ball but did not determine the spin. This requires one more piece of information which is provided by the constraint that, during the kick, the foot and the surface of the ball move together. From figure 10.14 we see that the tangential component of the foot velocity is $v_{\mathrm{f}}\sin\theta$, which for small angles is $v_{\mathrm{f}}\theta$. The surface velocity of the ball is the

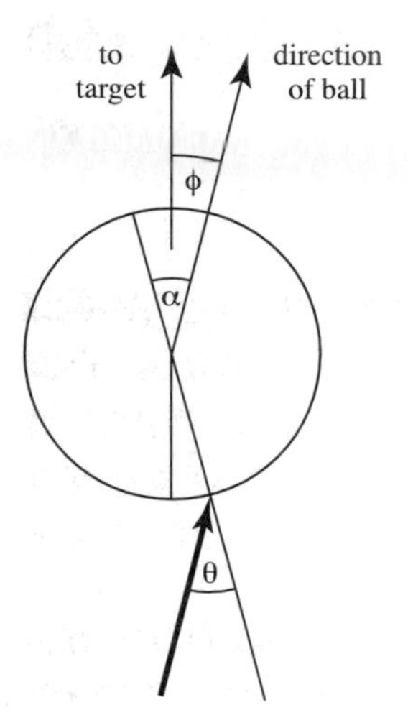

Figure 10.14. Showing the angle of the kick.

sum of the ball's sideways velocity and the surface rotation velocity, that is $u + \omega a$. Equating these velocities

$$u + \omega a = v_{\mathrm{f}}\theta.$$

This equation together with equation (66) gives both ω and u in terms of the controlled variables v_{f} and θ

$$\omega = \frac{3}{5}\frac{v_{\mathrm{f}}}{a}\theta \qquad \text{and} \qquad u = \frac{2}{5}v_{\mathrm{f}}\theta. \tag{70}$$

The angle α can now be determined using equations (69) and (70) to obtain

$$\alpha = \frac{2}{5(1+e)}\theta. \tag{71}$$

The dependence of θ and α on ϕ comes from equations (65), (68), (70) and (71) which give

$$\theta = \frac{20(1+e)\ell}{3C_{\mathrm{s}}L}\phi \tag{72}$$

$$\alpha = \frac{8\ell}{3C_{\mathrm{s}}L}\phi. \tag{73}$$

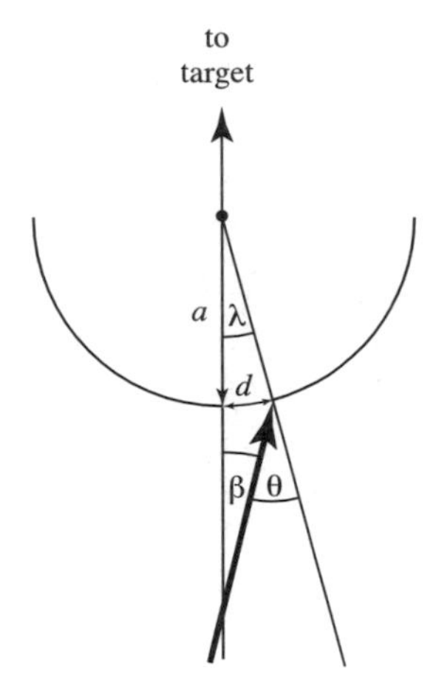

Figure 10.15. Introducing the angles β and λ.

(v) Complete prescription for kick

Figure 10.15 defines the problem. We want the direction of the ball to be at an angle ϕ, and we need to know the angle β of the kick and the off-centre distance, *d*, of its placement. It is seen that $d = \lambda a$, and so the problem reduces to that of finding the angles λ and β which produce the angle ϕ required for the ball to end up on target.

From figure 10.16 it is seen that the angles are related by

$$\lambda = \alpha - \phi$$

and

$$\beta = \theta - \lambda = \theta - \alpha + \phi.$$

Using equations (72) and (73) for θ and α gives λ and β in terms of ϕ and, recalling that $\phi = D/L$ and $\lambda = d/a$, we obtain the final requirements on the placement and the angle of the kick to give a displacement, *D*, of the flight over a distance *L*

$$\frac{d}{a} = \left(\frac{8\ell}{3C_s L} - 1\right)\frac{D}{L}$$

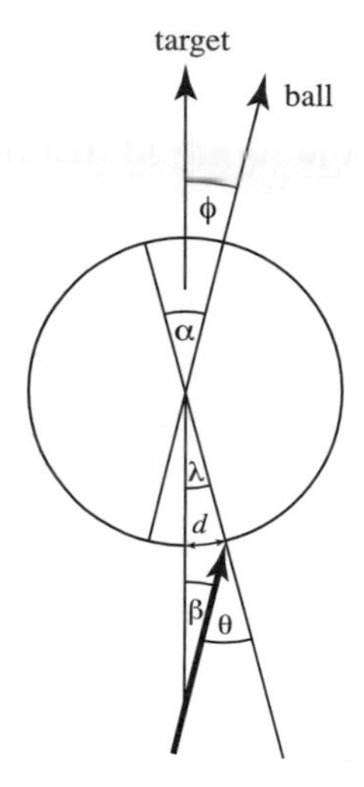

Figure 10.16. The full geometry of the kick.

and

$$\beta = \left(1 + \frac{4(3+5e)\ell}{3C_sL}\right)\frac{D}{L}.$$

Using the numerical values $m = 0.43\,\mathrm{kg}$, $\rho = 1.2\,\mathrm{kg\,m^{-3}}$ and $A = 0.039\,\mathrm{m}$ gives $\ell = 9.2\,\mathrm{m}$. As explained earlier we do not have an accurate value for C_s but a reasonable estimate is 0.5. Substituting these values with $e = 0.5$ we obtain

$$\frac{d}{a} = \left(\frac{49}{L} - 1\right)\frac{D}{L}$$

and

$$\beta = \left(1 + \frac{135}{L}\right)\frac{D}{L} \quad \text{radians}$$

$$= 57\left(1 + \frac{135}{L}\right)\frac{D}{L} \quad \text{degrees.}$$

It is interesting that d can be of either sign although with the value of C_s used it will almost always be positive. For a

25 m kick with a displacement D of 1 m the angle, β, of the kick to the target line is 15°.

The distance d is the required distance of the kick on the ball from the target line. The distance from the line through the ball in the direction of flight is greater. It is seen from figure 10.16 that this is given by the angle α, the distance on the surface being αa, and from equation (73)

$$\alpha a = \frac{8a\ell D}{3C_s L^2}.$$

With the numerical values used above and the ball radius $a = 0.11$ m

$$\alpha a = 5.4\frac{D}{L^2} \quad \text{metres}$$

so that for a kick with $L = 25$ m and $D = 1$ m the distance from the centre-line along the line of flight is about a centimetre.

5.1. Probability of scoring

If the ratio of the scoring rate of the stronger team to that of the weaker team is R, the probability, p, that the next goal will be scored by the stronger team is $R/(R+1)$ and the probability for the weaker team is $1 - p = 1/(R+1)$.

If one goal is scored in a match, the probability that it is scored by the stronger team is p and by the weaker team is $1 - p$. If there are N goals in the match the probability that they are all scored by the stronger team is p^N. The probability that the weaker team scores all the goals is $(1-p)^N$.

The probability, P, that the stronger team scores n goals out of N is

$$P = \frac{N!}{n!(N-n)!}\, p^n (1-p)^{N-n}$$

where N factorial is defined by

$$N! = N(N-1)(N-2)\cdots 1$$

and similarly

$$n! = n(n-1)(n-2)\cdots 1$$

and $0! = 1$.

6.1. Probability of scoring *n* goals in time *t*

For a team with a scoring rate of r goals per hour probability of scoring n goals in time t, measured in hours, is

$$P = \frac{(rt)^n}{n!}\,e^{-rt}. \tag{74}$$

where

$$e = \frac{1}{0!} + \frac{1}{1!} + \frac{1}{2!} + \frac{1}{3!} + \cdots = 2.718\cdots$$

and P has a maximum at $t = n/r$ given by

$$P_{max} = \frac{n^n}{n!}\,e^{-n}.$$

6.2. Probability of the score (*n*, *m*)

If teams 1 and 2 have scoring rates of r_1 and r_2 the probability that team 1 has scored n goals and team 2 has scored m goals in time t is, from equation (74),

$$P_{n,m} = \frac{(r_1 t)^n (r_2 t)^m}{n!m!}\,e^{-(r_1+r_2)t}.$$

6.3. Probability of scoring first in time *t*

The probability that a team has not scored ($n = 0$) in a time t is given by equation (74). Noting that $(r_1 t)^0 = 1$ and $0! = 1$

we obtain

$$P_0 = e^{-rt}.$$

If the scoring rates for teams 1 and 2 are r_1 and r_2 the probability that neither team has scored is

$$P_{00} = e^{-(r_1+r_2)t}.$$

The probability that team 1 scores in dt is $r_1\,dt$ and so the probability that neither team has scored at time t and team 1 scores in dt is

$$dP_1 = e^{-(r_1+r_2)t} r_1\,dt$$

and integrating from $t = 0$ gives the probability that, in a time t, team 1 has scored first

$$P_1 = \frac{r_1}{r_1 + r_2}\left(1 - e^{-(r_1+r_2)t}\right).$$

It is seen that P_1 rises from 0 at $t = 0$ to a limit of $r_1/(r_1 + r_2)$.

6.4. Random motion

Random motion can be treated theoretically by taking averages over time. The movement of the ball around the pitch does not allow a thorough theoretical description but a rough model is perhaps of interest.

It is quite usual on television to be given the percentage of the time which the ball has spent in parts of the pitch. For example, the length of the pitch is often divided into three parts and the percentage given for each part. For a theoretical model the pitch can be divided into many more parts and in the limit to an infinite number of parts. Choosing a sufficiently long time to obtain a satisfactory average we can then draw a graph of the distribution of the ball over the length, x, along the pitch. Such a graph is illustrated in figure 10.17 for a pitch of length 100 m. f is called the distribution function which can be measured in seconds per metre. The behaviour

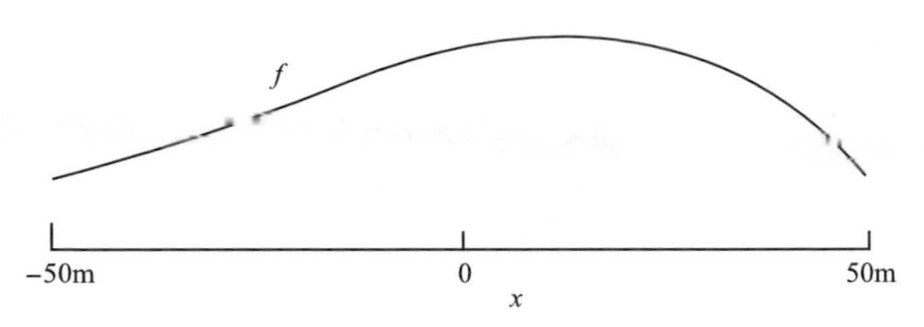

Figure 10.17. An example of the distribution, f, of the ball's time averaged position along the pitch.

of f for random motion can be described by the differential equation

$$\frac{\partial f}{\partial t} = \frac{\partial}{\partial x}\left(D(x)\,\frac{\partial f}{\partial x}\right)$$

where D, the diffusion coefficient, depends on x. The steady solution of this equation ($\partial f/\partial t = 0$) would be f = constant. The fact that f is not a constant arises from the strength and deployment over the pitch of the teams' resources. It is difficult to measure this precisely but it can be represented in the equation by a term $C(x)\,\partial f/\partial x$ to give

$$\frac{\partial f}{\partial t} = C(x)\,\frac{\partial f}{\partial x} + \frac{\partial}{\partial x}\left(D(x)\,\frac{\partial f}{\partial x}\right).$$

This equation is called the Fokker–Planck equation. The steady state is now described by

$$C(x)\,\frac{\partial f}{\partial x} + \frac{\partial}{\partial x}\left(D(x)\,\frac{\partial f}{\partial x}\right) = 0.$$

In practice we expect the 'steady' solution to evolve during the match principally due to change in $C(x)$.

6.5. Intercepting a pass

We calculate here the criteria for the interception of a pass made along the ground, directly toward the receiving player.

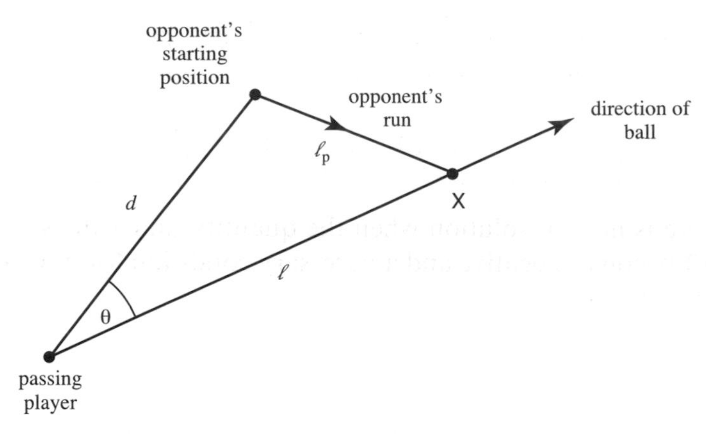

Figure 10.18. Geometry of the interception calculation.

The geometry is shown in figure 10.18. It is clearly a necessary condition for interception that the intercepting player must be able to reach some point on the ball's path before the ball reaches that point. We therefore need to calculate the time, t_b, for the ball to reach any point X, a distance ℓ along the ball's path, and the time, t_p, for an intercepting player to reach the same point. A successful interception requires that $t_p \leq t_b$ for some position of X, that is for some distance ℓ.

If the speed of the ball is s_b the time to reach X is

$$t_b = \frac{\ell}{s_b}. \tag{75}$$

Taking the speed of the player to be s_p, he can reach X in a time

$$t_p = \frac{\ell_p}{s_p}. \tag{76}$$

From the geometry ℓ_p is related to the separation, d, of the two players and the angle θ by

$$\ell_p^2 = d^2 + \ell^2 - 2d\ell \cos\theta. \tag{77}$$

For interception $t_p \leq t_b$ and the limits of interception are therefore at $t_p = t_b$, so that from equations (75), (76) and (77)

$$s_p^2 \ell^2 = s_b^2 (d^2 + \ell^2 - 2d\ell \cos\theta).$$

This is a quadratic equation for the limiting ℓ, and interception is possible for any ℓ between the two solutions

$$\ell = \frac{d}{1-(s_p/s_b)^2}\left[\cos\theta \pm \left(\left(\frac{s_p}{s_b}\right)^2 - \sin^2\theta\right)^{1/2}\right]. \qquad (78)$$

There is no real solution when the quantity under the square root becomes negative and a necessary condition for interception is therefore

$$\frac{s_p}{s_b} > \sin\theta.$$

This condition is necessary but not sufficient because there are two situations where the receiving player can intervene. Figure 10.19 illustrates the possibilities.

In the first case the receiving player is between the passer and the earliest point of interception. If ℓ_r is the distance between the passer and the receiver, the condition for the receiver to intervene is

$$\ell_r < \ell_{min}$$

where ℓ_{min} is the smallest interception length given by equation (78)

$$\ell_{min} = \frac{d}{1-(s_p/s_b)^2}\left[\cos\theta - \left(\left(\frac{s_p}{s_b}\right)^2 - \sin^2\theta\right)^{1/2}\right].$$

In the second case the receiving player must be able to run to a position $\ell \leq \ell_{min}$ in the time taken for the opponent to reach ℓ_{min}. From equations (76) and (77) this time is

$$t_{pm} = \frac{(d^2 + \ell_{min}^2 - 2d\ell_{min}\cos\theta)^{1/2}}{s_p}. \qquad (79)$$

If the receiving player starts at a distance L from the passer and runs at a speed s_r, his time to reach ℓ_{min} is

$$t_{rm} = \frac{(L - \ell_{min})}{s_r}. \qquad (80)$$

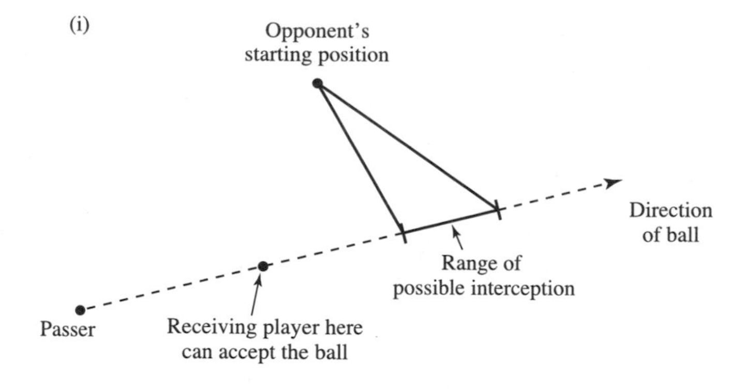

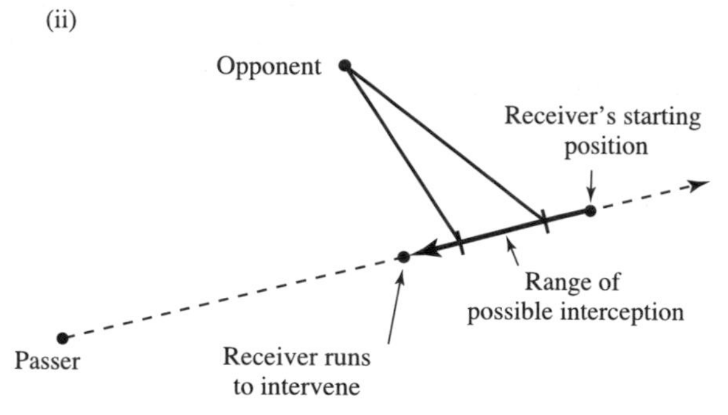

Figure 10.19. (i) Receiving player takes a short pass which the opponent cannot intercept. (ii) Receiving player runs to prevent interception.

Using the equations (79) and (80), the condition for successful interception by the receiving player, $t_{\mathrm{rm}} < t_{\mathrm{pm}}$, becomes

$$(L - \ell_{\min}) < \frac{s_{\mathrm{r}}}{s_{\mathrm{p}}}(d^2 + \ell_{\min}^2 - 2d\ell_{\min}\cos\theta)^{1/2}.$$

7.1. Spread in league points

The spread of points in a final league table has two contributions. The first arises from the random effects in each team's performances and the second is due to the spread of abilities among the league's teams.

In statistical theory the spread is measured by the so-called standard deviation. If a quantity x has a set of N values labelled x_n and the average value is $\bar{x}_n$, the standard deviation, σ, is defined as the square root of the mean of the squares of $x_n - \bar{x}_n$, that is

$$\sigma = \left(\frac{1}{N}\sum_n (x_n - \bar{x}_n)^2\right)^{1/2}.$$

We can use a simple model to estimate the spread in teams' points totals arising from the random variations of each team's results. The spread due to teams' differing abilities can be eliminated by taking all the teams to be equal. We then take reasonable probabilities for match results, $\frac{3}{8}$ each for a win and a defeat and $\frac{1}{4}$ for a draw. If each team plays N matches there will, on average, be $\frac{3}{8}N$ wins, $\frac{3}{8}N$ defeats and $\frac{1}{4}N$ draws. If there are 3 points for a win, 1 for a draw and 0 for a defeat the average number of points per game will be

$$\bar{P} = \tfrac{3}{8}3 + \tfrac{1}{4}1 + \tfrac{3}{8}0 = \tfrac{11}{8} \quad \text{points}$$

and the expected standard deviation over N games is then

$$\begin{aligned}\sigma &= \left(\tfrac{3}{8}N(3 - \tfrac{11}{8})^2 + \tfrac{3}{8}N(\tfrac{11}{8})^2 + \tfrac{1}{4}N(1 - \tfrac{11}{8})^2\right)^{1/2} \\ &= 1.32N^{1/2} \quad \text{points}\end{aligned}$$

For $N = 38$, as in the Premiership, the standard deviation would be 8.1 points.

We can now examine the actual standard deviation of points obtained by teams in the Premiership using the final league tables. Averaging over five years this turns out to be $\sigma = 13.6$ points. The extra spread in points over the basic value 8.1 can be attributed to the spread in abilities of the Premiership teams. Figure 10.20 gives a graph comparing the spread in points due to randomness alone with that actually obtained in the Premiership.

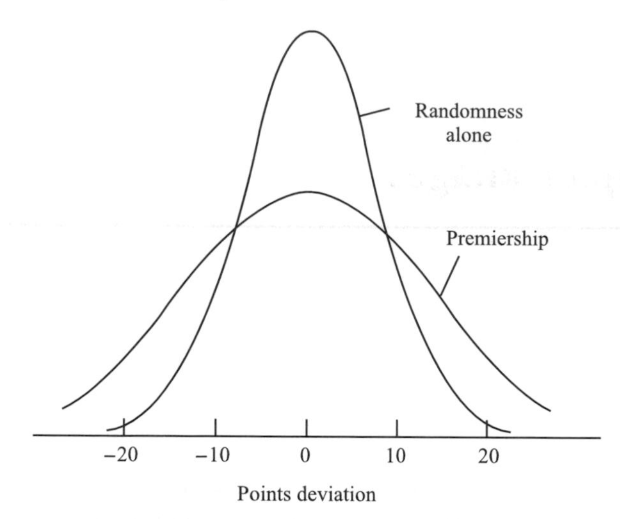

Figure 10.20. Graph of the distribution of points about the mean for randomness alone and for the Premiership.

It is clear that the random element plays a large part in determining a team's final points total and can therefore influence which team becomes champion. The discussion about the 'best team' in chapter 7 is an attempt to quantify this.

Chapter images

1. Selected frames from high speed (4500 frames/sec) photography of a bounce (*D. Goodall*). The ball moves from left to right and the bounce is seen to make the ball rotate.
2. Powerful kick by Ruud van Nistelrooy of Holland. (*Photograph by Matthew Impey*, © *Colorsport*.)
3. Oliver Khan of Bayern Munich jumps to catch the ball. (*Photograph by Andrew Cowie*, © *Colorsport*.)
4. Boundary layer separation in the wake of a circular cylinder.
5. Referee Mike Pike showing firmness. (*Photograph by Matthew Impey*, © *Colorsport*.)
6. 'The Thinker' by Auguste Rodin. (© *Photick/Superstock*.)
7. The first League table. Preston were undefeated in this season and also won the F.A. Cup.
8. England's World Cup winning team, 1966. Captain Bobby Moore holds aloft the Jules Rimet Trophy. (© *Popperfoto/PPP*.)
9. Professional football's cash flows.

10. Newton's Laws of Motion, from the Principia

    Law I. Every body perseveres in its state of rest, or uniform motion in a straight line, except in so far as it is compelled to change that state by forces impressed on it.

    Law II. Change of motion is proportional to the motive force impressed, and takes place along the straight line in which that force acts.

    Law III. Any action is always opposed by an equal reaction, the mutual actions of two bodies are always equal and act in opposite directions.

Bibliography

Although ball games have probably been played for thousands of years the basic scientific ideas which underlie the behaviour of balls only arose in the seventeenth century. Galileo was the first to discover the rules governing the flight of projectiles and calculated their parabolic trajectory.

The greatest step was made by Isaac Newton with his *Mathematical Principles of Natural Philosophy* (London, 1687) – usually called *The Principia*. In this magnificent book he proclaimed the basic laws of mechanics – the famous three Laws of Motion and the Law of Gravity. *The Principia* is available in a recent translation by I. B. Cohen and Anne Whitman (University of California Press, 1999).

It is a sign of Newton's versatility that in this book he also addresses the problem of the drag on a sphere moving through a medium. Although his model was not valid, it enabled him to discover the scaling of the drag force. He found the force to vary as $\rho A V^2$ as is now used in the equation $F = \frac{1}{2} C_D \rho A V^2$ (given in Chapter 10, section 4.2).

When we come to the Magnus effect, it is remarkable that the first recorded observation of the effect is due to Newton. He had noticed that the flight of a tennis ball is affected by spin. In the *Philosophical Transactions* of the Royal Society of London (1672) he recalls that he 'had often seen a Tennis ball, struck with an oblique Racket, describe such a curve line' and offers the explanation. 'For a circular as well as a progressive motion being communicated by that stroak, its parts on that side where the motions conspire, must press and beat the contiguous Air more violently than on the other, and there exert a reluctance and reaction of the Air proportionally greater.' In 1742 Benjamin Robins published his treatise *New Principles of Gunnery* and reported his observations of the transverse curvature of the trajectory of musket balls. He stated that its 'Cause is doubtless a whirling Motion acquired by the Bullet about its Axis' through uneven rubbing against the barrel (pages 91–93). A later edition gives details of his experiments. Subsequently Gustav Magnus observed the effect on a rotating cylinder mounted in an air flow in an investigation of the deflection of spinning shells. His paper 'On the deviation of projectiles, and on a remarkable phenomenon of rotating bodies' was published in the Memoirs of the Berlin Academy in 1852 and in an English translation in 1853.

The real understanding of drag and the Magnus–Robins effect awaited the discovery by Ludwig Prandtl of the 'boundary layer'. He described the concept in the Proceedings of the 3rd International Mathematical Congress, Heidelberg (1904). The classic text on boundary layers is *Boundary Layer Theory* by Hermann Schlichting, first published in German in 1951 and then in English by McGraw-Hill. There are many books on fluid mechanics: a clear modern text is *Fundamentals of Fluid Mechanics* by Munson, Young and Okiishi (Wiley).

For those wishing to study the derivation of the probability formulas an account is given in the excellent book *Probability Theory and its Applications* by Feller (Wiley).

Turning to books more directly relevant to the Science of Football, first mention must go to *The Physics of Ball Games* by C. B. Daish (Hodder and Stoughton) which, unfortunately, is now out of print. This book concentrates somewhat on golf, and only briefly deals with football. However, it is a good introduction to the underlying physics. A book which would appear from its title to be more closely related to the present one is *Science and Soccer* (Spon), edited by Thomas Reilly. However, the content of this book is quite different and more practical, dealing with subjects such as physiology, medicine and coaching.

Index